Design Considerations for the Multi-stage Space Elevator

International Space Elevator Consortium
Spring 2019

Authors:
John M. Knapman
Peter Glaskowsky, Dan Gleeson, Vern Hall,
Dennis Wright, Michael Fitzgerald, Peter Swan

Design Considerations for the Multi-stage Space Elevator

Copyright © 2019 by:

John M. Knapman

Published by Lulu.com

pete.swan@isec.org

978-0-359-33232-8

Printed in the United States of America

International Space Elevator Consortium

Preface

The International Space Elevator Consortium vision is to have:

"A World with inexpensive, safe, routine, and efficient
access to space for the benefit of all mankind."

One of the beautiful aspects of working in future engineering concepts and architectures is that you stretch out into the unknown. The challenges that are facing space elevator development are real, and sometimes intimidating. However, the team that has assembled to take on these challenges is quite remarkable. Our chief architect is laying out a preliminary baseline with technological challenges identified with verification and validation testing shown for each major segment. In addition, alternative approaches are presented so that the development team can assess where it should head. Many of these potential approaches are to enable technologies and designs that could lead to earlier development.

A Multi-stage Space Elevator concept has been developed out of several known technologies with experimentation and operational experiences. This concept has the potential of enabling space elevators earlier than previous plans while relying on known transportation technologies and approaches. This study report lays an alternative for Space Elevators at the doorstep of our future access to space. We are encouraged about the future of Space Elevators as we believe we are:

Closer than Most People Think.

Signed: *Peter A. Swan, Ph.D.*
           President ISEC

Executive Summary

The study team took on the challenge of expanding the body of knowledge pertaining to the Multi-stage Space Elevator.

To build a space elevator, the toughest challenge is to find material that is strong enough for a self-supporting tether. Building it in multiple stages is a way of overcoming that challenge. Using the concept of dynamically supported structures, it is possible to build upwards from the earth's surface and provide supports for the lowest parts of the tether, where gravity is strongest. A five-stage design would support a tether made of carbon fiber yarn that is commercially available today. A two-stage design can support a tether with less than one-third of the strength previously thought necessary.

The known technologies of magnetic levitation and evacuated tubes are required together with some recently developed techniques for maintaining stability and supporting continuous operation. A substantial submarine (or underground) structure is required, but the capacity and operations of climbers are similar to those previously proposed. The overall mass of the tether also remains much the same. Provision has been made for dealing with space debris.

Table of Contents

International Space Elevator Consortium

1 Introduction

This study report will parallel the previous International Space Elevator Consortium study reports with a year-long study assessing the idea of a multi-stage Space Elevator. The initial chapter lays out the historical setting and then shows the breakout across the chapters. Several volunteers worked on this report and contributed some ideas that were widely discussed previously but not recorded. Many of the concepts are unique and new. The brainstorming started early, was orchestrated during the 2017 International Space Elevator Conference mini-workshop on the topic and continued until the last word was finalized. This series of topics focusing on multiple stages was open to serious engineering considerations and major definitional activities. The topics had been discussed before the study was initiated, but there remained many puzzles to address and answers to propose.

The lead author, John Knapman - Ph.D., has developed this idea over the last eight years with feedback from the space elevator team. After this introductory chapter, there are chapters dealing with many details required to initiate developmental concepts: the strategic approach for space elevators, refinement of the lower stage segments, and assessment of individual arenas leading to a totally integrated series of lower segments with a space elevator architecture. Finally, the conclusions are presented concisely with additional information in the appendices.

1.1 Principal Challenge

The principal challenge in building a space elevator is to find a material strong, long and light enough for the tether, so that it can reach down to the earth's equator from an altitude of 100,000 km. There appear to be two ways of addressing this problem:

1. Continue research into materials such as carbon nanotubes (CNTs) that have shown immense strength in tiny lengths.

2. Investigate ways of bypassing the problem by building structures upwards from the earth's surface.

By pursuing both lines of research in parallel, we maximize the chances of being able to build a space elevator sooner rather than later.

Very large funds have been spent on research into strong materials, and work continues [1]. The challenges are great, but the rewards would also be great, as there are many terrestrial applications for strong materials, for example constructing longer bridges or building aircraft.

Building structures upwards from Earth's surface has received relatively little interest, although there has been some work done. Notable are the space fountain and the launch loop [2][3]. A development of the launch loop called High Stage One has been proposed for the space elevator to deal with winds and ice in the atmosphere [4].

High Stage One of the space elevator is held up by fast-traveling objects called bolts, by analogy with the bolts fired by a crossbow. To minimize friction, they travel inside evacuated tubes; magnetic levitation is used to prevent them touching the sides of the tubes. To save power, permanent magnets provide the required force; they are stabilized by electromagnets. A technique called active curvature control enables High Stage One to maintain stability in the presence of gusting winds [5]. It provides a light-weight mechanism and improves on an earlier proposal [6].

High Stage One forms an arch between two surface stations, whereas the multi-stage space elevator uses similar techniques but in a vertical form. In that respect it is more like the space fountain, but it uses radically different methods. The space fountain used induction coils to support vertical tubes, but the multi-stage space elevator uses the more efficient magnetic levitation method.

1.2   Meet in the Middle

A good way of looking at progress in development of strong, long, light materials is to see how far they can reach down towards Earth from the geosynchronous altitude (GEO) of 35,850 km. The standard model as defined in the IAA study requires a specific strength of 38 MYuri.*0 On a reasonable set of assumptions, that reaches down into the atmosphere. A material with only 11 MYuri specific strength can reach down to an altitude of 6000 km. To reach further down without exceeding the acceptable mass budget, the tether needs to be supported. The reach of other materials is shown in Table 1, ending with Torayca carbon fiber yarn†, which can reach to 14,600 km and is commercially available.

---

* The Yuri is the tensile strength in Pascal (Pa) divided by the mass density in kg/m³. A MYuri is GPa/g.cm⁻³.
† Torayca is a trademark of the Toray Corporation of Japan

International Space Elevator Consortium

Table 1 Using weaker materials by adding more stages

Material Strength MYuri	Number of Stages	Total Mass of Tether Metric Tons	Altitude of Top Stage Km
11	2	6600	6000
7.3	3	6600	9700
5	4	6400	12,900
3.9 (Torayca)	5	7200	14,600

As we make progress with the multi-stage space elevator and with materials, we look to a convergence so that the upward reach of the vertical structure meets the downward reach of the tether (Figure 1).

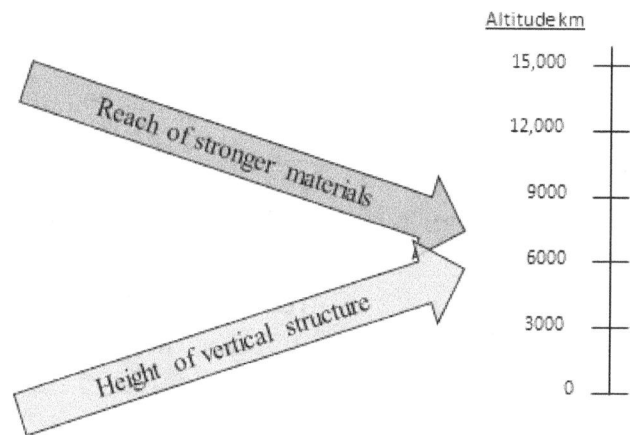

Figure 1 Technology convergence

1.3   ISEC Study Process:

The yearly study process has been consistently investigating elements of the space elevator for the last nine years. The ISEC has a process to pick topics and then conduct a year-long in-depth analysis on critical topics for Space Elevator development. This focus enables the ISEC to prioritize activities and leverage volunteers with expertise in the chosen fields. The single focus on a topic for a particular year enables the community to bring its strengths together and address the topic at the yearly conference, inside the organization's journal, CLIMB, and magazine, Via Ad Astra, and through the study process with a resulting report. The topics chosen by the Board of Directors of ISEC have been:

2010 – Space Elevator Survivability, Space Debris Mitigation

2011 – Carbon Nanotube Developmental Status
2012 – Space Elevator Concept of Operations
2013 – Design Considerations for the Tether Climber
2014 – Space Elevator Architecture and Roadmaps
2015 – Design Considerations for the Earth Port
2016 – Design Considerations for the Space Elevator GEO Node, Apex Anchor
2016 – Design Considerations for the Space Elevator Communications
   Architecture.
2017 – Design Considerations for Space Elevator Simulation
2018 – Design Considerations for Multi-Stage Space Elevators

Each study goes through a similar process, such as:

August 2017	ISEC selects topic at Board of Directors meeting
	Topic announced at the yearly conference
Aug-Dec 2017	Team formed and initial outline of study topics
Jan-Mar 2018	Specific items discussed, analyzed and studied
Mar-Aug 2018	Paper topics submitted to the ISEC International Conference
August 2018	Focus at space elevator conference on topic
	Mini-workshop brainstorming provides feedback. [Appendix C]
Sep-Jan 2018	Study topics drafted as chapters in the report
Jan-Feb 2019	ISEC Review of Final Document
Feb-Mar 2019	Final review with top level peer review
April 2019	Publish Study Report

This report will be available on the ISEC website in hardback form for sale and as pdf, for free at www.isec.org

1.4 Chapter Layout

One of the keys to ISEC's study success is that the work is accomplished by a diverse set of space elevator enthusiasts with special skill sets related to the study topic. Each study is accomplished in about a year with the objective of having the report available for everyone at the conference the following year. The report is laid out as follows:

Chapter 2 Background: This chapter lays out the motivations for the concept in view of current progress in materials research. It looks at the feasibility condition and shows how sensitive the result is to seemingly small changes in material strength. It also outlines some of the design choices available.

Chapter 3 Principles behind the Multi-Stage Space Elevator: This chapter explains how the structure will work, how a dynamically supported structure is possible and what its main components are.

Chapter 4 Dealing with the technological challenges: The greatest challenge is space debris. The chapter also covers matters related to high speed, stability and minimizing power consumption.

Chapter 5 Earth Port: This chapter recognizes that some changes to the Earth Port will be needed compared to the more conventional approach, but these are kept to a minimum.

Chapter 6 Design of the Upper Stages: This chapter explains how each of the upper stages can support the tether down to the next stage below.

Chapter 7 Development program with cost estimates for prototype: This chapter proposes a program that requires funding.

Chapter 8 Conclusions and Recommendations: This chapter lays out the conclusions and recommendations from the study members. The consensus of the study team is laid out so that the recommendations can be initiated and the conclusions clearly understood.

> **Appendix A** is the ISEC Vision and Mission
> **Appendix B** is the Acronym List and Terminology List
> **Appendix C** is the Architecture Engineering Baseline Change Management
> **Appendix D** is the minutes from the brainstorming sessions
> **Appendix E** is the mathematical support

2 Background

2.1 Recent work on materials

Individual CNTs have been shown to have a strength of 200 GPa or about 150 MYuri.0 However, combining CNTs into a ribbon is much harder. A good result is that of Wang at el., who achieved 9.6 GPa with a density of 1.85 g/cc, which amounts to 5.2 MYuri (Figure 2) [9].

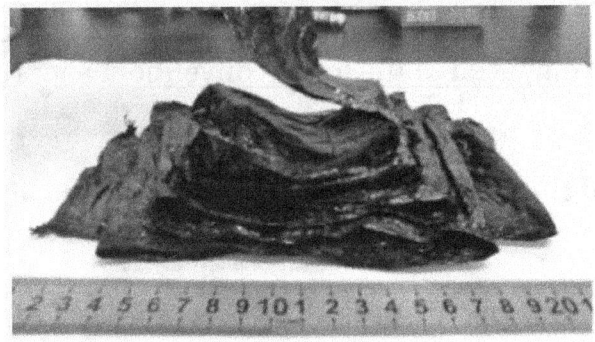

Figure 2 A length of CNT ribbon with a specific strength of 5.2 MYuri

2.2 Proposals on material strength

Edwards assumed a strength of 130 GPa and a density of 1.3 g/cc, leading to a specific strength of 100 MYuri [10]. Obayashi Corporation assume 150 GPa with density 1.3, leading to a specific strength of 115 MYuri [11].

Shelef showed that lower strengths are possible [12]. He developed the feasibility condition to allow us to trade off the material strength against the mass and taper ratio of the tether.‡ Using this condition, the IAA study selected a specific strength of 38 MYuri with a taper ratio of 1:6 and a safety margin of 40%.

2.3 The feasibility condition

The feasibility condition states that the space elevator must be able to lift enough material to replace that lost due to damage and wear. In addition, during construction it must be able to lift its own weight in a reasonable time. This is because the construction process requires a thin tether to be launched to GEO by rocket and then used to lift more material until the full tether mass is ready for operations. The key measure is the time it takes to double the tether's mass. If the

---

‡ The taper ratio is the ratio between the mass per unit length of the tether at GEO and at its base.

International Space Elevator Consortium

complete tether mass is 6400 tonnes[§] and we launch 100 tonnes by rocket, we need six doublings to complete construction. If the doubling time is 18 months, construction takes nine years.

The multi-stage space elevator is rather different. Only the tether above the top stage has to have this doubling capability; the lower tethers are hauled up from earth with the ambits during construction. However, the requirement to be able to replace losses due to damage and wear is much the same.

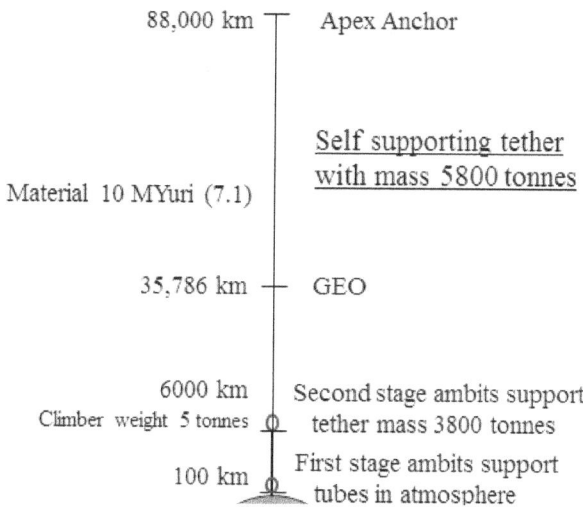

Figure 3 Architecture using 10 MYuri material

We know from Chapter 3 of the IAA study that a tether with a mass of about 6600 tonnes has the required properties, provided its strength is 38 MYuri. To play it safe, we can require that the total tether mass over all stages of the tower is no more than 6600 tonnes, or we can take advantage of the less stringent requirement below the top stage to allow a higher total. Figure 3 shows a 10 MYuri material reaching down to 6000 km with a two-stage structure below that. Two is the minimum number of stages, the first stage to support the elevator in the atmosphere and the second stage much higher in space. The mass above the second stage is 5800 tonnes; the mass between stages one and two is 3800 tonnes. Interestingly, increasing the strength by just one to 11 MYuri reduces these masses to 4200 and 2400 tonnes respectively, giving a total mass of 6600 tonnes, which satisfies the more stringent version of the feasibility condition.

Table 1 shows how we can trade off material strength with the number of stages, on the assumption that we want to obey the more stringent feasibility condition. In the

---

[§] One tonne is one metric ton, equal to 1000 kg.

calculations, the strengths shown are divided by 1.4 to provide a safety margin of 40%.

Torayca is commercially available. Torayca with five stages is feasible if we allow an overall tether mass of 7200 tonnes (Figure 4). A possibly beneficial approach would be to double the tether mass below the second stage from 1360 to 2720 tonnes. That would increase the overall tether mass, but it would double the capacity up to the second stage, allowing us to build up a substantial reserve of tether material and repair climbers at the second stage. They could be used to service the tether above the second stage, which has a total mass of 5840 tonnes and satisfies the feasibility condition. A similar logic could be applied to other configurations, allowing a relaxation in the strength requirement or in the number of stages.

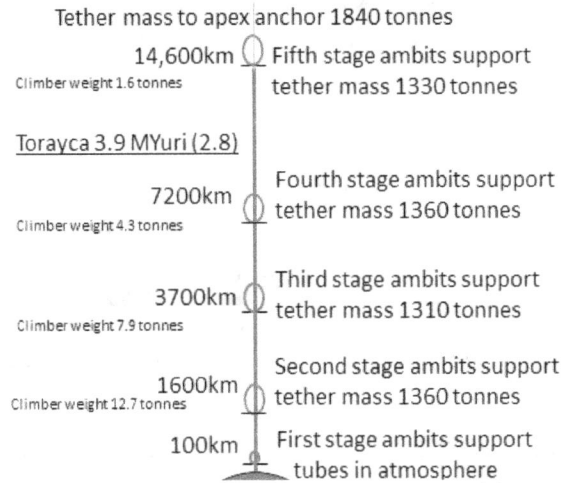

Tether mass to apex anchor 1840 tonnes

14,600km — Fifth stage ambits support tether mass 1330 tonnes
Climber weight 1.6 tonnes

Torayca 3.9 MYuri (2.8)

7200km — Fourth stage ambits support tether mass 1360 tonnes
Climber weight 4.3 tonnes

3700km — Third stage ambits support tether mass 1310 tonnes
Climber weight 7.9 tonnes

1600km — Second stage ambits support tether mass 1360 tonnes
Climber weight 12.7 tonnes

100km — First stage ambits support tubes in atmosphere

Figure 4 Configuration using commercially available Torayca yarn

2.4   Climbers on the multi-stage space elevator

Two kinds of climber are needed to gain the maximum benefit from the multi-stage space elevator, *tube climbers* in the atmosphere and tether climbers above the atmosphere. This is to deal effectively with winds, ice and electrical storms in the atmosphere. In space, the hazards are different, namely, space debris, meteoroids and radiation.

Tube climbers are used to ascend to the first stage at 100 km high. They draw power from the tubes, which also support their weight, so they can travel at any time of day or night. Their climbing mechanism is designed specifically to ascend the tubes. A tube climber carries a tether climber along with its payload and protects it from the atmosphere, somewhat like the box protection method that was previously proposed.0 At the first stage, equipment operated remotely from the earth's surface takes the tether climber from the tube climber and transfers it to the tether ready for its ascent. The tether climber deploys its solar panels or laser power receiver and starts the climb at an appropriate time. The tube climber can return to Earth when it is operationally convenient.

The tether extends from the first stage (at 100 km) all the way through the geosynchronous altitude to the apex anchor. The second and higher stages provide intermediate supports to the tether to allow it to be made of material that is less strong than was previously thought necessary. When a tether climber arrives at one of these upper stages, it has to pass over the supports that hold the tether up. To minimize the weight of the tether climber, the mechanism to do this should be built in to the supports. A good way to achieve this is to provide two supports for the tether spaced 30 meters apart, as in Figure 5. The mechanism removes the lower support so that the weight of the tether (and climbers) hangs on the upper support. The tether climber then passes that point and parks while the mechanism restores the lower support to its proper place. Then it removes the upper support to allow the tether climber to pass and continue its ascent. Finally, it restores the upper support.

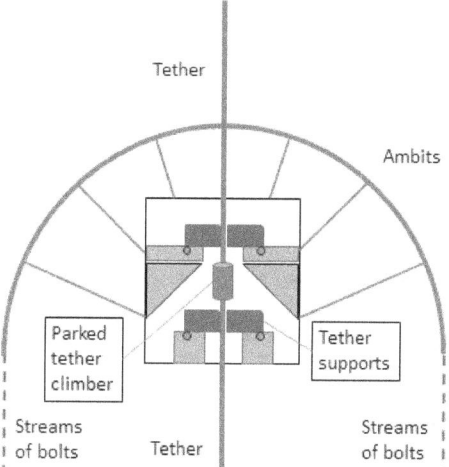

Figure 5 At the second or higher stage, a tether climber is parked while the supports are moved

2.5   Maintenance

Normal operations at high altitude are to be performed by automatic machines or remotely controlled machines. This includes routine safety and reliability checks. Human visits to the first stage (100 km) will be rare and will require space suits. Human visits to the second or higher stages will be even rarer, since the mechanisms are deliberately designed to be simple and robust with high degrees of automation.

The bolts travel in space, but they return to Earth approximately every hour, when safety checks can take place and faulty bolts extracted.

2.6   Alternatives considered

1. It would be possible to extend the tether right down to the surface, but the tether would then be subject to winds, ice and electrical storms. Recent studies have shown that the effects of winds on the tether in the atmosphere would be severe in times of rough weather; to mitigate those effects a different shape of tether – narrower than the one meter width needed in space – would be needed, and so a different climber mechanism would be required anyway compared to that needed in space [4]. Since there will already be tubes in the atmosphere that are stabilized against wind and have ice and lightning mitigation measures, we may as well use them for climbing.

2. We could winch payloads or complete tether climbers (suitably protected from the atmosphere) up to the first stage at 100 km. This would avoid the need for a special climbing mechanism in tube climbers, but it would be necessary to find a way to protect the winch's cable from wind and other atmospheric damage.

3. Instead of lifting a complete tether climber, a tube climber could lift the payload in a container. Then remotely controlled machines at the first stage would move the payload in the container on to a tether climber. That would mean that tether climbers are held at the platform on the edge of space, and so the facilities at the first stage would need to be more sophisticated. Tube climbers would still have to lift and complete tether climbers from time to time for management purposes, so this solution is more complex overall.

4. In the second and higher stages, an alternative to having the movable tether supports would be to separate the upper part of the tether from the lower part. When the tether climber arrives at the stage (Figure 6) it waits for an automated mechanism to transfer it from one part of the tether to the next (Figure 7). Separating the parts of the tether this way may make maintenance easier, particularly at the altitudes where space debris is most problematic, i.e., below 2000 km.

International Space Elevator Consortium

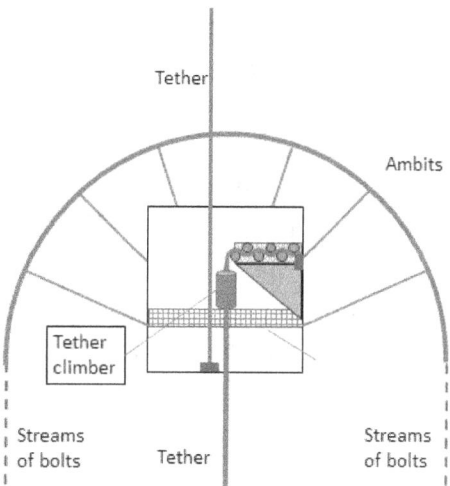

Figure 6 A tether climber awaiting transfer from the lower to the upper part of the tether

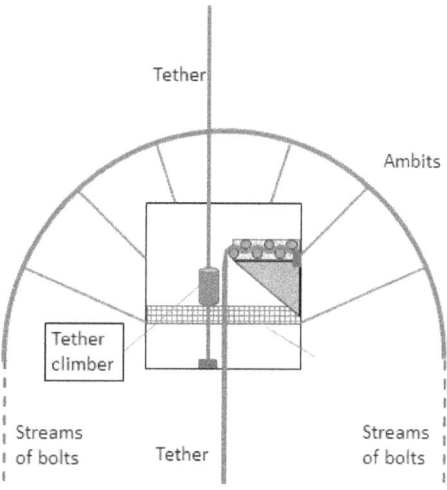

Figure 7 The tether climber just after the transfer

2.7 The ride

First you need to get to Kiribati in the central Pacific Ocean, probably via Hawaii. As you approach, you will see a cluster of tubes rising from the terminus platform. The platform is floating and is stabilized in the same way as an oil exploration platform. When preparations are complete, you will board a tube climber and enter the capsule inside it. This capsule will be your accommodation for a week as you travel to an altitude of 35,786 km.

The tube climber will draw power from the tubes as it ascends through the atmosphere to the edge of space, a journey of about an hour. As you climb higher, the sky will darken and more and more stars will become visible. However, your weight will not reduce noticeably, since you are not in orbit. The ride will be as gentle as riding an elevator in a tall building.

The edge of space, known as the Kármán line, is 100 km up. There you will arrive at the first stage, where automatic equipment will transfer you inside your capsule from the tube climber to a waiting tether climber. You will see the tether rising into the distance above you. You can just make out the streams of small objects called bolts, which are rising above you in parallel lines. Because they are so fast and there are gaps between them, you will see them as a transparent blur. Normally, there is no-one here, but you may just see an engineer doing some routine checks and maintenance.

Your solar-powered tether climber will commence its ascent at dawn to make the most of the sunlight. The acceleration is very gentle, taking four minutes to reach its speed of 75 km/h. As it climbs, it gets further away from Earth, the gravity becomes less, and it loses weight. Therefore, it can climb faster, reaching 185 km/h at a height of 1600 km by the end of the day. The next day, its speed gradually increases to 200 km/h as it climbs to 4600 km. The blur of the streams of bolts is still visible about 800 meters away on each side of the climber.

At about 9 am on the third day, we reach the second stage, 6000 km above Earth, where we slow down and stop for a couple of minutes. You can see the streams of bolts reaching the semi-circular shape known as the ambit. They hold the second stage up; it turns them around and sends them back down. In turn, the ambit holds up the equipment at the second stage, and it supports the tether. The tether just below the second stage is thicker than it is above, although you may not be able to see the difference. There is a mechanism on the second stage to transfer the climber from the lower part of the tether to the upper part.

By this time, you are much lighter. Your weight is just over a quarter of what it was on Earth. The same applies to the climber. As a result, it accelerates to its speed of 280 km/h and continues to accelerate gently to its maximum speed of 300 km/h. The climber continues at this speed, with stops due to darkness depending on the season. On the seventh day, it reaches the GEO node at 35,786 km. Here, you and the climber are weightless. The GEO node is undergoing extensive construction work to turn it into a small city where people can live and work. It is a gateway to the solar system and, ultimately, the galaxy and beyond. You can still see the tether continuing further away from Earth. Earth itself is by now a ball looking much more distant, although it is still the largest object in the sky by far. It no longer seems to be down; in microgravity, 'up' and 'down' have little meaning.

3 The principles behind the multi-stage space elevator

3.1 Introduction

There have been ideas proposed about how to construct a tower that reaches into space. The space fountain would rise to a great height supported by fast projectiles inside a tube [2]. The projectiles pass through a tube containing electric induction coils that create drag on the projectiles, causing an equal and opposite lifting force on the tube and on any payload it carries. The problem is that the power consumption of the coils is too great, approximately equal to the average power consumption of a medium-size country. More promising is an adaptation of High Stage One (Figure 8), which has already been proposed for the space elevator as a method of dealing with Earth's turbulent atmosphere (Chapter 5 in ref. [7]).

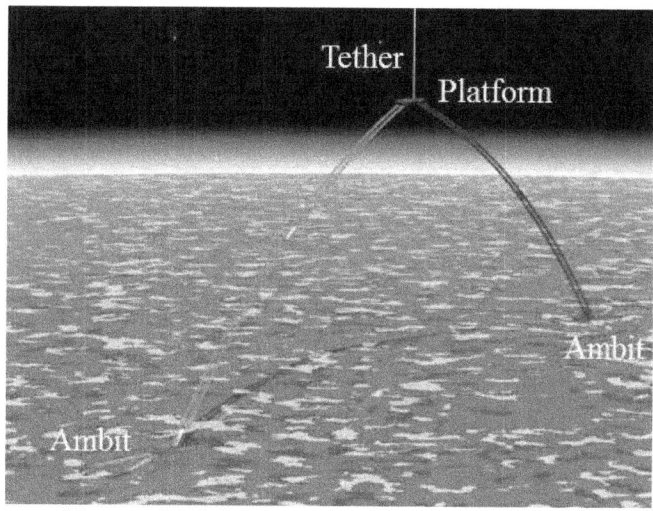

Figure 8 High Stage One of the space elevator

3.2 High Stage One

High Stage One uses permanent magnets for lift. Electromagnetic coils are used only for stabilization, and electronic controls with rapid response times keep the power required to a minimum. The bolts travel inside evacuated tubes. They rise to a height of 40 km, although the original Launch Loop was proposed to reach 80 km. A key limiting factor on height is the need to avoid space debris. The bolts circulate in tubes between two horseshoe-shaped structures called *ambits*, which are similar to each other. The bolts are designed to be mass produced in a factory to minimize costs and maximize reliability.

International Space Elevator Consortium

To build a tower, the idea is to have one large ambit at the earth's surface and several smaller ones at very high altitudes, so that the bolts travel up from the surface to the upper ambits, where they are turned around, creating lift. This lift is used to support the part of the tether below the ambits. The descending bolts return to the surface and are turned around by the lower ambit. Their kinetic energy is conserved. Thrusters in the lower ambit boost the bolts' speed to make up for any residual losses due to friction. The bolts store their electrical energy in capacitors with battery backup, which they recharge while passing through the lower ambit.

An important advantage of High Stage One is its ability to deal with winds and ice in the atmosphere by passing the forces down to the surface rather than up the tether. The tower will still be able to use this ability in the atmosphere. In space, the design is different: there is no need for tubes at all, as friction in the vacuum of space is minimal.

3.3 The underlying physics

A bolt rising under gravity will gradually lose momentum. If v_0 is its vertical velocity at the earth's surface, h is its height, g is the acceleration due to gravity at the surface, and R is the earth's radius, the vertical velocity v satisfies

$$\frac{dv}{dt} = -g\frac{R^2}{(h+R)^2} + \frac{v_l{}^2}{h+R}$$

Here, v_l is a bolt's lateral velocity, i.e., in the orbital direction. The space elevator rotates with the earth, and so v_l increases linearly with height. A descending bolt is subject to the opposite acceleration, so that it arrives at the lower ambit on the earth's surface at much the same speed as it departed. Ascending and descending bolts exchange lateral momentum through their magnetic connection, which creates a Coriolis force between them.

The upper ambits will reverse the bolts' vertical velocity. We consider a stream of bolts with mass m per meter. Note that m varies with speed because the bolts get closer together as they slow down, according to Bernoulli's principle. In fact, $m = m_0 v_0/v$, where m_0 is the mass per meter at the earth's surface. Every second, a mass mv of bolts arrives at the upper ambit, where it is turned around. Consequently, the rate of change of momentum due to a stream of bolts at the upper ambit is $2mv^2$. This is the vertical force on the upper ambit, and it has to support the structure associated with the upper ambit, the tether beneath it down to the next lower stage, and any climbers that may be on the tether.

International Space Elevator Consortium

On this basis, we can calculate that a bolt leaving the lower ambit at the equator needs a vertical velocity of 7.85 km/s to reach 6000 km at a vertical velocity of 1.2 km/sec, where it turns around by passing through the second-stage ambit in the two-stage design. The bolts are spaced 1 m apart, and the bolt mass is 0.6 kg, so the stream of bolts has a mass per unit length of 0.6 kg/m. At 6000 km, the spacing is 15 cm and the stream has a mass of 4 kg/m. The upward force $2mv^2$ is equal to 11.5 Mega Newtons (MN). This is enough to support the weight of the second stage and the weight of the tether and climbers.

The upper ambit is semicircular, and its expected diameter is 1 km. Its mass is estimated at 10 kg/m, making a total mass of 31 metric tons. Then its weight will be 83 kN, bearing in mind the reduced gravity at 6000 km. The tether and climbers weigh just under 9 MN, so there is plenty of spare capacity. However, the bolt speed at the second stage is very sensitive to the bolt speed at the surface. Reducing their surface speed from 7.85 to 7.80 km/s reduces their speed at the second stage from 1200 to 800 m/s. This in turn halves the levitation force provided by the ambit.

It is possible to extract power from the moving bolts by retarding their motion a little. This could be used to power a laser or microwave transmitter, for example. However, the estimates for the tether and climbers assume a worse case that climbers are solar powered and are limited to 4 MW, which means that they travel slowly near the earth and get faster as their weight decreases. It also means that they rest at night.

The worst case using solar power occurs at dawn, when a climber commences its ascent from the first stage (at 100 km altitude). That climber is subject to nearly full gravity – its weight is 190 kN reduced by only 3% – based on a climber mass of 20 metric tons. Another climber has parked overnight at 1370 km altitude, where dawn is earlier than on the surface. It has had time to climb to 1580 km by the time dawn reaches the surface at the equator at the worst time of year, which is the equinox. At 1580 km, its weight is 125 kN. At the same moment, a third climber is at 4600 km where its weight is 65 kN. The mass of the tether needed to support these weights (plus its own weight) is 2400 metric tons, assuming a specific strength of 11 MYuri and a working specific strength of 7.9 MYuri, allowing the same 40% safety margin as in the standard model. The taper ratio is 47:1. Its weight is 8.5 MN, bringing the total weight of tether plus climbers to just under 9 MN.

The height of the second stage was chosen to ensure that the overall tether mass was the same as that in the standard model, namely about 6600 metric tons. The tether above 6000 km extends to 88,000 km, and its mass is 4200 metric tons. Its taper ration is 13:1. This overall mass was chosen to satisfy the feasibility condition, which allows enough capacity in the space elevator for maintenance and repair while still providing a useful capacity for profitable payloads. The apex anchor mass is 840 metric tons. Similar calculations for the three-stage, four-stage and five-stage versions are summarized below.

	Height (km)	Mass of tether above (metric tons)
Stage One	100	2400
Stage Two	6000	4200
Apex Anchor	88,000	Apex anchor mass 840
Total tether mass		**6600**

Table 2 Summary of two-stage space elevator with specific strength 11 MYuri

	Height (km)	Mass of tether above (metric tons)
Stage One	100	1400
Stage Two	1530	1320
Stage Three	3700	3870
Apex Anchor	82,000	Apex anchor mass 500
Total tether mass		**6590**

Table 3 Summary of three-stage space elevator with specific strength 7.3 MYuri

	Height (km)	Mass of tether above (metric tons)
Stage One	100	1400
Stage Two	2000	1400
Stage Three	5350	1500
Stage Four	12,900	2100
Apex Anchor	77,000	Apex anchor mass 180
Total tether mass		**6400**

Table 4 Summary of four-stage space elevator with specific strength 5 MYuri

	Height (km)	Mass of tether above (metric tons)
Stage One	100	1360
Stage Two	1530	1310
Stage Three	3700	1360
Stage Four	7200	1330
Stage Five	14,600	1840
Apex Anchor	72,000	Apex anchor mass 165
Total tether mass		**7200**

Table 5 Summary of five-stage space elevator with specific strength 3.9 MYuri (Torayca carbon fiber yarn)

The choices of where to place the stages were made to optimize the specific strength for a given number of stages within the constraint of keeping the overall tether mass to 6600 metric tons, apart from the five-stage example, where we have stretched the condition to embrace a material that is already available commercially. The method employed was a manual manipulation of spreadsheets to come up with approximately equal tether masses between stages so as to minimize the overall tether mass. Further optimization is possible.

Stability

Maintaining stability of a dynamically supported structure requires new approaches compared to more conventional structures. In the atmosphere, tubes are subject to winds and sometimes ice and electric storms. Provisions for ice and lightning just add some weight, but winds will cause unpredictable instabilities and must be dealt with carefully.

The straightforward way to deal with winds is to use guy wires up most of the height to 100 km. This adds considerably to the overall weight compared to a more sophisticated technique called *active curvature control* [5]. It exploits the way that a tube will naturally bend in the wind, and it propagates and limits the movement so that a centrifugal force in the tube opposes the wind. The centrifugal force is mv^2/r for a radius of curvature r, and so faster bolts need less curvature than slower bolts. The algorithm uses electronically controlled electromagnets between the bolts and the tube. At least three electromagnets operate on a bolt in each of the two orthogonal directions. They exploit the comparative rigidity of the bolts compared with the tubes. The reference contains implementation details and a mathematical proof of stability.

In space, the considerations are different. There is, of course, no wind. The forces are tidal and are predictable. There are no tubes, and so the only controllable factor once the bolts have left an ambit is the force between ascending and descending bolts. Appendix E contains a description of the method, a mathematical proof of its stability, and a description of a simulation that has confirmed the results.

Tube material

Using active curvature control means that the bending properties of the tubes are less important than if we use guy wires over the full length of the tubes. However, the tubes will be under considerable tension, since they are supported from above. In addition, it is vital to maintain the integrity of the vacuum.

Aluminum is an excellent material to minimize outgassing. It is suitable as a thin inner tube, or possibly two concentric tubes to maximize vacuum integrity, surrounded by a reinforcing polymer to give high tensile strength. Carbon-Kevlar composite tubing is a suitable material.

3.4 Dynamically supported structures

In the bolts, it is important to eliminate eddy currents as far as possible and to minimize hysteresis losses. In addition, a low power consumption is required. Using non-conducting magnets such as ceramic ferrites is the easiest way to avoid eddy currents. They have only one third of the field strength of state-of-the-art sintered Neodymium-Iron-Boron (generally just called neodymium) magnets, which implies that considerably more mass is needed in each bolt compared with using neodymium. Making the bolts more massive is actually beneficial, however, because it increases their lifting capacity at the upper ambits. If it is decided to use neodymium, it is possible to minimize eddy currents by building magnet blocks out of 1×1 mm tubes of neodymium thinly coated with an insulator.

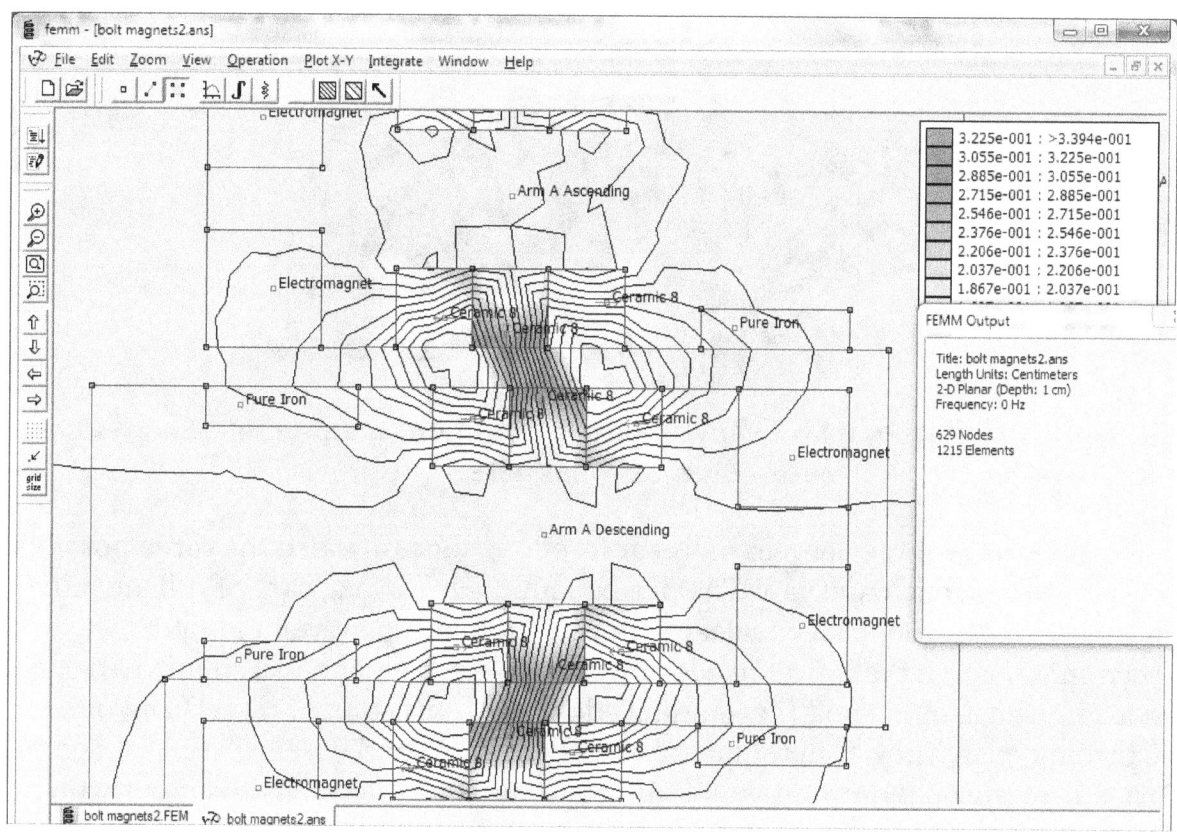

Figure 9 Permanent magnets in Halbach arrays

The arrangement of the permanent magnets is based on the Halbach array and illustrated in Figure 9. ** Originally, Halbach arrays were circular, but a simplified linear array works well, concentrating the field on one side and creating a strong attraction when the magnets have opposite poles. There is a face-to-face attraction in the y-direction, but they are in opposite directions on each side of the bolt, thus canceling each other out. The electromagnets have the job of keeping the bolt in a central position so that the balance is maintained. There is a healthy lateral restoring force in the x-direction that pulls the magnet arrays into alignment opposite each other. The two sides of the bolt reinforce each other. This is the force that holds the bolt in line with the tracks in the tubes and in the ambits. The direction of travel is the z-direction into or out of the paper in Figure 9.

** Produced using FEMM – Finite Element Method Magnetics by David Meeker at http://www.femm.info/wlkl/HomePage

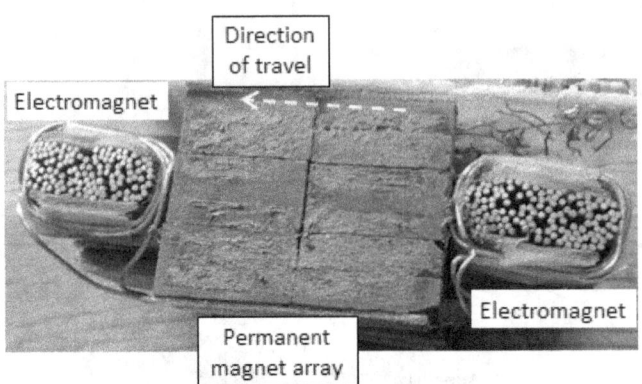

Figure 10 Part of a bolt showing the permanent magnet array and the electromagnets

In addition to the three degrees of linear freedom, there are also the three possible rotations of roll, pitch and yaw. The permanent magnets take care of roll and pitch, but the electromagnets have to deal with yaw, as that is a movement in the y-direction. It means that a bolt needs at least two electromagnets (Figure 10), one in the leading edge of the bolt (i.e., leading edge in the direction of travel) and one in the trailing edge. They each require a pair of position sensors (Figure 11), one on each side of the bolt. Infrared diodes are suitable as position sensors: subtracting the signals from each side of the bolt gives a reliable measurement of the centrality of the position, from which the electronic controls determine whether to pull in the positive or negative y-direction.

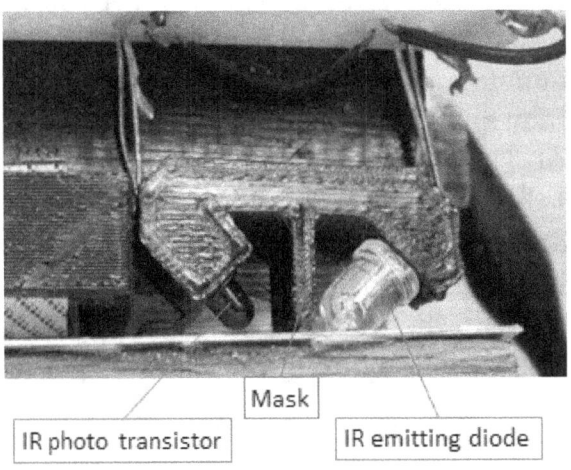

Figure 11 A position sensor

The core of each electromagnet could be made of a non-conducting ferrite material but, for an alternating current in the windings, it is usual to build the core from thin insulated sheets of iron or steel. It is even more effective to build the core from strands of insulated iron wire less than 1 mm in diameter, running from end to end

International Space Elevator Consortium

of the electromagnet. A superior material for the electromagnet cores is Metglas,[††] which has much better hysteresis properties, but it is available in thin sheets rather than wires. It should be possible to negotiate a special order of cores made of insulated wires for large quantities, although perhaps not for small-scale prototypes.

The coils of the electromagnets are wound round the cores conventionally, which has the effect that any currents induced on one side cancel those induced on the other side.

There will inevitably be some losses due to hysteresis in the iron cores of the electromagnets. When the coil is no longer energized, some residual magnetism remains. This may lead to some induced eddy currents, but these can be avoided by using multiple strands of insulated iron wire instead of steel plates. The prototype bolt illustrated in Figures 10 and 11 was built for High Stage One but is similar to the tube bolts needed for the multi-stage space elevator in the atmosphere, where the magnets in the bolts engage with a continuous magnetic track in the tubes. The free bolts that travel in space use the same principles, but they have two arms instead of one. Figure 12 shows a tube bolt traveling in a tube, which has been rendered transparent for the sake of visibility.

The electronics should respond in times below 1 µs. The faster the response, the less current needs to be passed through the coils of the electromagnets. Early prototypes will use general circuits with programmable chips, but later versions will use custom analogue chips that maximize response times and minimize power consumption.

Figure 12 A bolt in a tube with magnetic tracks

[††] Metglas is produced by Metglas, Inc.; see www.metglas.com

3.5 Dealing with friction and other losses

Magnetic levitation avoids friction, but some losses due to hysteresis and eddy currents are inevitable. In addition, despite attempts to minimize the power consumption of the electronics and electromagnets, the capacitors and batteries in the bolts will require recharging. One method of doing this is to provide additional coils on each bolt specifically designed to generate electric current as they pass by permanent magnets in the ambits and possibly in the tubes and in other bolts. This makes the overall system resilient even when there is a failure of power to the thrusters, because it allows bolts to keep going for a considerable time without thrusters. It is possible to design electronic controls in each bolt that make the decision as to when and where recharging should take place. In those bolts that travel in space without tubes, small solar panels could be incorporated.

Evacuating the tubes minimizes the losses due to air resistance, although there will inevitably be some residual air due to outgassing and occasional leaks. Thrusters in the lower ambit, and possibly elsewhere, have the job of making up for these losses by boosting the velocity of bolts as they pass. The thrusters can accelerate the bolts slightly beyond the speed required to ascend to their target altitude (e.g., 6000 km), and the bolts can generate power as they pass other parts of the ambit, although that will create a drag force that retards their speed. To produce 10 Watts of power over the 30 minutes it takes for a bolt to rise from Earth's surface to 6000 km requires 20 kJ of energy. The energy stored in a 500-gram bolt travelling at 7850 m/sec is 15.406 GJ. Even increasing the speed by just 1 m/sec to 7851 increases the energy to 15.410 GJ, an increase of 4 MJ, which is 200 times that required.

This amounts to a huge reservoir of stored energy, allowing the bolts to keep going long after a power failure and giving ample time for the restoration of normal service.

3.6 Magnetic levitation

A prototype bolt for High Stage One has been built and is shown in Figure 10 and Figure 11. The tube bolts used in the multi-stage space elevator are similar. Figure 13 shows the main components. The permanent magnets are arranged in a Halbach array. There is a Halbach array on each side of the bolt.

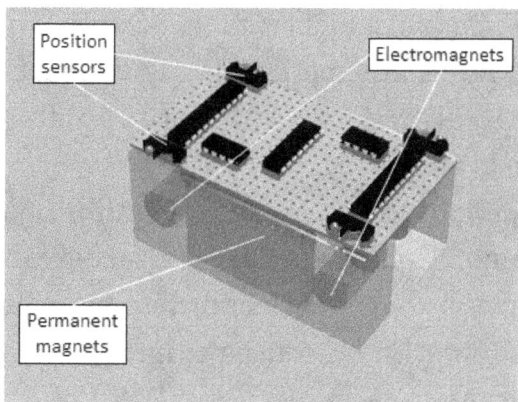

Figure 13 A tube bolt

A free bolt has two arms instead of one, as in Figure 14. Its components are similar.

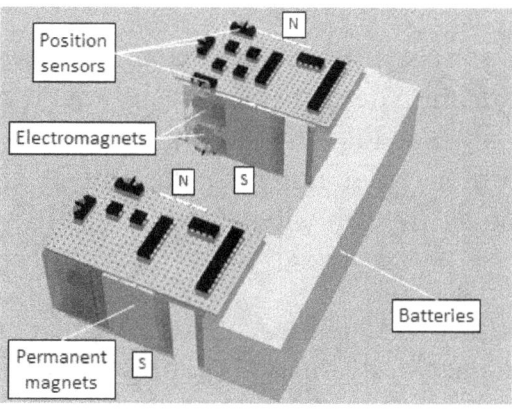

Figure 14 A free bolt

A tube bolt has the following major elements. Doubtless these details will change when we gain some experience from practical tests.

1. A 3D printed framework holds three permanent magnets on each side, which match the permanent magnets in the track along which it is designed to travel. In a vacuum, 3D printed material may lead to outgassing; another material may be needed.

2. Two electromagnets, one at each end, reach across the width of the bolt. When energized, each electromagnet simultaneously repels on one side and attracts on the other; in this way, a small movement of the bolt away from the center of the track can rapidly be corrected. When a bolt is at rest, it may be in contact with the permanent magnets on one side of the track; the electromagnets are powerful enough to move it to the center of the track in

International Space Elevator Consortium

this worst case, which typically occurs at start-up. To achieve this, fairly large currents of tens of Amps are needed for a few milliseconds.

3. Four 4.7 mF (4.7×10⁻³ Farad) capacitors are used to provide the surge in current for the electromagnets.

4. The bolt's power supply comes from three Lithium-ion batteries connected in series, giving a nominal voltage of 11.1 – in fact it varies up to 12.6 V.

5. There are four sensors, two at each end, detecting the distance from the bolt to the track at each side. These consist of infrared-emitting diodes with infrared-sensing transistors.

6. Electronic controls monitor the output from these transistors, subtracting those on opposite sides and setting appropriate levels for the power transistors that allow current to flow through the electromagnets.

Permanent magnets

The permanent magnets are ceramic ferrites, each 10×10×40 mm with the direction of magnetization along a 10 mm depth. They are arranged as an array of three similarly to a Halbach array (see Figure 9), with the central magnet's north pole facing the track and the other two magnets' north poles facing the central magnet. In the track and magnets' poles are opposite so that they attract the bolts.

Electromagnets

The electromagnets have a core of 125 parallel iron wires, which have an insulating coating of enamel to minimize eddy currents. The iron wires are 50 mm long and 1 mm in diameter. The resultant core measures approximately 9×14×50 mm. The insulated copper windings are in four layers of 0.71 mm diameter, making 280 turns in all. A smaller electromagnet would require thinner winding wire. For example, to achieve consistent scaling a half-size electromagnet would need double the current density (measured in Amperes per square meter), which can be achieved by making the wires $1/\sqrt{2}$ times the diameter, requiring winding wire of 0.5 mm diameter. The required diameter is $d' = d\sqrt{f}$ for a scaling factor of f.

There is a trade-off needed here. In general, a stronger electromagnet has a higher inductance. This means it is more sluggish in responding to a signal. Hysteresis and other magnetic properties of the core are also important in minimizing losses. However, inductance is key, as it governs the time taken for the current in the electromagnet (and hence the magnetic field and the restoring force) to respond to an applied voltage. During this time, the bolt will continue to move to one side of the track, and so more energy is needed in the electromagnet to bring it back to the central position of unstable equilibrium. The inductance of the present electromagnets is 2.1 mH, and they respond to an applied current in about 7 μsec. A smaller electromagnet will have lower inductance and so will respond more rapidly, giving an overall reduction in energy use.

Another way to reduce the inductance is to split each electromagnet into two with a gap in between. In the present prototype, this could be achieved by having two electromagnets each 20 mm long with a 10 mm gap between them. The magnetic force is affected very little, but each electromagnet has less than half the inductance. Connecting them in parallel halves the inductance again but doubles the current that has to be supplied. The net effect of connecting them in parallel should be a beneficial reduction in the energy consumed, because the faster response should enable a more rapid restoration of equilibrium.

Surge capacitors

The 4.7 mF 16 V capacitors are quite bulky. They are only needed at start-up, because the lithium-ion batteries are unable to deliver the required surge in current. A current as high as 30 Amps is generated for a few milliseconds. More experiments are needed to work out the minimum required capacitance. One way to minimize the needed current surge may be to start by activating one end of the bolt, getting it into a temporary quasi-stable position, allowing half a second for the capacitors to recharge and then activating the other end. This approach may reduce the overall capacitance required.

Scaling the bolt size by a linear factor f reduces the magnetic forces by f^2. However, the current required would be the same to satisfy the requirement for a factor f increase in the current density. To overcome this consideration, we would need to make a further increase in the number of winding turns in the electromagnets. One solution is to add additional windings to each electromagnet using small-gauge wire (e.g., 0.1 mm) in a separate circuit used only at start-up. Although this solution is more complicated, it is likely to be much more efficient than the present approach and could even eliminate the need for surge capacitors altogether.

Batteries

The sensors and digital integrated circuit (IC) require a 3 V power supply, but 12 V is used for the electromagnets to maximize the response time. A step-down voltage regulator IC is used to provide the 3 V supply. There is not enough experience to determine what battery capacity is really required. It is possible to use a step-up voltage regulator in conjunction with a surge capacitor to produce 12 V from a battery with a smaller voltage.

The present battery pack consists of three lithium-ion camera batteries and measures 40×35×24 mm. For scaling down, we would like to be able to use a smaller battery pack, but the available technology dictates a diameter of 25 mm upwards. As with a mobile phone, the battery size is likely to be the chief determiner of overall size.

So far, no on-board recharging facilities have been implemented. We simply remove the batteries to recharge them.

Sensors

Early experiments used capacitive sensors to measure the distance between a bolt and the sides of the tube. A 40 MHz oscillating signal in the bolt was supplied to plates of metal foil opposite metal-foil plates in the tubes. They formed a low-pass filter that attenuated the signal depending on the spacing between the bolt and the tube. These proved difficult to implement, mainly because of the technical difficulty of forming reliable high-frequency electrical connections with the plates in the bolt.

A more successful technique has been to arrange a mask between infrared-emitting diodes and infrared-sensitive transistors in the bolts so that the reflection from metal foil plates in the tubes strikes the sensor in rough proportion to the distance.

A third technique that may be worth trying is to place magneto-resistors in the bolts so that they detect the field strength due to the proximity of the magnets in the tubes. This technique could lend itself more effectively to miniaturization. A challenge might be to avoid disturbance from the magnetic field of the electromagnet.

In all cases, the difference between the measurements on each side is used to calculate the required voltage across the electromagnet.

Electronic controls

Each of the two electromagnets – one at each end of the bolt – has its own sensors and control circuits. The central component is a digital processor on a chip, which includes analogue-to-digital (AD) converters, a digital-to-analogue (DA) output, and a set of binary on-off outputs. It is programmed from a laptop computer, which also has access to monitoring and debugging information. Once programming and testing are complete, the computer can be disconnected.

Input from the sensors passes through two AD converters, which are each sampled at the rate of 2 MHz. The processor is programmed to subtract the results from each side of the bolt and use a table for efficient conversion to the output voltage. It calculates the velocity at which the bolt is moving sideways and uses this to apply damping to avoid excessive oscillations. The result is passed to the DA output to drive the power transistors that control the voltage across the electromagnet. An operational amplifier (opamp) is used to deliver the current required to drive the power transistors and to convert the voltage range; the processor delivers a voltage in the range 0 to 1.5 V, whereas the power circuits operate at 0 to 12 V.

The circuit consists of a number of resistors and integrated circuits wired on a board with holes spaced 0.1 inch apart. An order-of-magnitude space saving could be achieved by using surface mount technology (SMT). In addition, it would be possible for a single IC measuring 3 to 4 millimeters to be manufactured encompassing the functions of the whole circuit except for the sensors and the power transistors. A single IC would be too expensive for prototyping, but the economies of scale for large numbers are immense.

Greater efficiency could be achieved by implementing the main feedback loop from sensors to electromagnets using analogue circuitry. For the sake of robustness, it makes sense to build the circuitry for the main feedback loop in triplicate with best-of-three logic to detect inconsistencies. The digital IC would still be useful for start-up, shutdown, monitoring and notifying the appropriate facility in the lower ambit to remove a malfunctioning bolt when inconsistencies are detected.

3.7   Alternatives to maglev, e.g., electrostatic levitation

In the vacuum of space, it may be possible to use electrostatics to maintain the position of bolts as they pass each other. For example, two plates forming a 50 pF capacitor with a gap of 1 mm can exert a force of 10 N if the potential difference is 20 kV. This is likely to be more efficient, with much lower losses, than using electromagnets.

4 Dealing with technological challenges

4.1   Introduction

Maglev trains are, of course, very much larger than the bolts used in the multi-stage space elevator, and the techniques needed are different in scale and detail, though not in principle. Magnetic bearings are used in high-speed pumps, and they also use electronic stabilization, so there is similarity there. However, our requirement is to be even more economical with power consumption while coping with a very high speed of travel. The key to this combination of challenges is to achieve the fastest possible response times whenever a deviation from the ideal position takes place.

4.2   Very high speed of travel

Although the speed requirements are demanding, it is instructive to compare with computer hard-drive disk technology that is already available. Our requirements are similar in terms of relative scale.

The outer track of a standard 7200 RPM 3.5" disk drive travels at 30 m/s with a clearance as little as 3 nm (10^{-9} meter). We design the clearance between the bolts and track to be 1 mm, and so the speed equivalent to that of a hard drive head is 10 km/s when comparing them scale for scale. Moreover, disks running at twice this speed are on the market. The maximum speed at which a bolt travels is 7.87 km/s, although the relative velocity of two bolts passing is twice that. Hence the requirements are not unprecedented, although the requirement to operate in a vacuum is different.

4.3   Minimizing power consumption

Speeding up response times in the process of electronic stabilization generally has the benefit of reducing the power requirement. With the popularity of battery-powered mobile devices, there is a world of experience in optimizing the power consumption of the electronics in the bolts. Examples include placing more functions into a single microchip and reducing the operating voltage. These and other ideas for optimization in the electronics can be explored as the opportunity arises.

4.4   Reliability

The reliability of electronics today is very high, and the progression towards incorporating more and more components into microchips is increasing their reliability as well as lowering their cost. For example, a four-layer chip can provide an electromagnet by aligning spiral windings in each layer [13]. Batteries can also be incorporated into chips as well as more conventional components such as resistors, capacitors and transistors.

International Space Elevator Consortium

Nevertheless, further measures will be required against failure. In particular, redundant circuits should be incorporated into bolts with routine, automated health checks much like those carried out on laptop computers. Devices could also be incorporated into the lower ambit to perform additional checks and to pick up any faults already detected by the on-bolt diagnostics. Failing bolts would then be removed from the lower ambit by being diverted into a graveyard channel.

4.5   Hazards in space

In space, the electronic components and circuits will be subject to radiation and heat. They will need some form of hardening to operate safely. Some work is needed in this area.

Debris

The greatest challenge in designing a method of supporting part of the space-elevator tether from the earth's surface is how to deal with space debris. The atmosphere shields the lowest part of the elevator from this bombardment, but we have to deal with it in space.

The space agencies have acquired a great deal of experience in shielding from the impact of human debris and meteoroids. Whereas debris typically travels at up to 14 km/s, natural meteoroids can have an impact velocity up to 70 km/s. The general philosophy is to mitigate the impact of non-trackable items up to 10 cm in size using shielding. Larger items are regularly tracked, and evasive action is taken.

The Whipple shield is a well-tried method.[‡‡] It is used on the International Space Station and on many other spacecraft and satellites. An outer aluminium layer causes hypervelocity objects to vaporize on impact, so that secondary objects are scattered sufficiently to spread their energy of impact on the inner layer; there is a gap of about 5 cm between these two aluminium layers. This gap may be vacuum, but it is usual to include thin layers of Nextel and Kevlar to absorb more of the energy before the secondary objects reach the inner layer. These shields have been developed to minimize mass while maximizing protection.

Whipple shield mass

The outer aluminium layer is typically about 2mm thick. The inner layer is about 4.8 mm, but this is for manned spacecraft, where puncturing the inner layer could be fatal. It is reasonable to assume a thinner inner layer of perhaps 3 mm with a

---

[‡‡] See the following web sites: http://orbitaldebris.jsc.nasa.gov/protect/shielding.html and http://www.esa.int/Our_Activities/Operations/Space_Debris/Hypervelocity_impacts_and_protecting_spacecraft

International Space Elevator Consortium

standoff between 5 and 15 cm. Hence the outer layer's mass is approximately $2600 \times 0.002 = 5.2$ kg/m², and the inner layer's mass is approximately 7.8 kg/m².

The tubes between 80 and 100 km altitude need a shield with a radius of 5 cm; the circumference is 31 cm, and the mass of the inner layer is 2.4 kg/m. The outer layer's radius is 10 cm, the circumference is 63 cm, and the mass is 3.3 kg/m. The combined mass of the shielding is 5.7 kg/m. To protect the upper ambits, a shield about double this size per meter is required.

Dealing with larger objects

To avoid tracked objects, the best method is to arrange for the streams of bolts to be parted so that there is a gap of several kilometers through which the dangerous object can pass harmlessly.

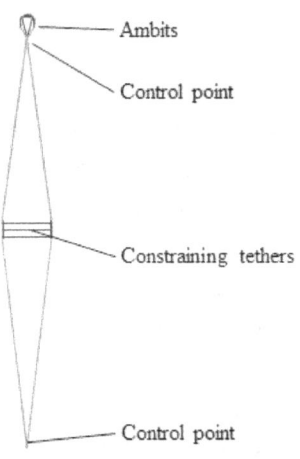

In Figure 15, one stream of bolts is separated from the other. Each of these streams consists of both ascending and descending bolts, which have to stay close so that the bolts emerging from an ambit are able to guide those bolts that are arriving and are about to enter. Another factor is that they have to compensate each other for the Coriolis forces, which are present because the tether's lateral velocity increases with increasing height. The Coriolis forces are caused by the orbital acceleration of bolts as they ascend and the matching orbital deceleration as they descend. At the earth's surface on the equator, they travel at the rotational speed of the earth, which is a horizontal velocity of 465 m/s. At 6000 km the tether and the bolts that support it travel at a horizontal suborbital velocity of 902 m/s. Hence the horizontal acceleration or deceleration is 0.073 m/s for every vertical km. Near the earth, the bolts that are to rise to 6000 km ascend at 7.8 km/s, and the horizontal acceleration is 0.7 m/s². The vertical velocity decreases to about 700 m/s near the upper ambit, at which point the horizontal acceleration diminishes to 0.05 m/s².

The operation of separating the streams of bolts is performed at two control points, one near or at the first stage ambits at about 100 km altitude, the other just below the second-stage ambits. These control points consist of a mechanism to turn the streams of bolts so that they diverge in the required direction, which is not

International Space Elevator Consortium

necessarily in the same plane as the ambits. Constraining tethers are placed a little over half way up from the lower control point. They cause the streams of bolts to converge instead of diverging. Turning the bolts requires inwards centripetal forces supplied by the constraining tethers. There are also relatively small vertical forces, and so vertical constraining tethers are needed too, as illustrated. By placing the constraining tethers more than half way up, we compensate for the slightly lower speed of the bolts higher up and provide a small net upward thrust sufficient to support the weight of the constraining tethers. When the threat is over, these tethers pull the tubes back together.

If the divergence is 1%, corresponding to an angle of 0.9°, and the lower control point is at 100 km altitude, the gap between the tubes at 200 km altitude is 1 km. At 3000 km altitude, the gap is a maximum of 32 km. If more than one stage is threatened by space debris, they will all have to undergo the same maneuver. The tether on which the climbers are ascending will stay on one side; the choice of which side is arbitrary.

If an ambit is under threat from space debris, it will have to be lowered a few kilometers. If the tether above the highest ambits needs to be moved, we can swing it from the top of the ambits and pull it back afterwards, much as was envisaged for High Stage One [4].

Dealing with smaller objects

Individual bolts in free space are expendable. If one or two are struck by a small piece of debris or a meteoroid, they will be deflected from the path of the other bolts and will be lost. This may be good enough for large items of debris as well; in that case we could avoid the necessity of special provisions for avoiding larger objects. However, in the normal course of travel when descending, free bolts must engage with the free bolts that are ascending in order to guide them to the ambit and to transfer lateral momentum from the descending to the ascending bolts (the Coriolis force). The bolts are spaced in proportion to their vertical velocity. As they rise, they slow down and get closer together. Consequently, at the highest velocities, free bolts encounter those traveling in the opposite direction briefly and intermittently. By contrast, when bolts travel in a tube and in an ambit, they are constantly engaged with the tracks of permanent magnets there, whether they are tube bolts or free bolts.

Collisions

Space debris is a significant hazard to the bolts traveling in free space. Occasionally, a meteoroid may also strike. The effects are minimized by having no mechanical bond between bolts, so that the impact is localized and does not affect the larger structure. Nevertheless, work is needed to examine to what extent cascades resulting from an impact may affect neighboring bolts.

A key factor is the choice of materials in a bolt. Permanent magnets and electromagnets will have to be made of suitable metal, but the structural material of the bolt may be chosen to minimize the effects. Nylon has been shown to be particularly well behaved in collisions at astronomical velocities.[§§] It vaporizes rather than producing dust or larger ejecta.

The detailed behavior depends on the relative size of the colliding objects. Studies have shown a clear distinction between the cases. An incoming object much larger than a bolt will simply smash cleanly through, carrying vapor, dust and other ejecta with it. An object much smaller than a bolt will itself be shattered while making a crater on the bolt. Vital electronics on the bolt therefore need to have triple redundancy so as to cope with disruption of this sort. A protective layer of nylon would also be desirable.

The most complicated case is a collision with an object that is similar in size to a bolt. Both objects will shatter or vaporize. We can analyze the resultant momentum change of the bolt and the debris. The bolt will have a vertical velocity of about 6 km/sec. The debris velocity is likely to be about 10 km/sec in an orbital direction. Conservation of momentum tells us that the center of mass of the resultant fragments and vapor takes on the combined momentum according to the vector equation:

$$(m_b + m_d)\boldsymbol{v}_r = m_b\boldsymbol{v}_b + m_d\boldsymbol{v}_d$$

Here, m_b is the mass of the bolt, $\boldsymbol{v}_b$ is its velocity vector, m_d is the mass of the debris, $\boldsymbol{v}_d$ is its velocity vector, and $\boldsymbol{v}_r$ is the resultant velocity vector. If the masses are equal and the velocities are orthogonal, then $\boldsymbol{v}_b = (0, v_b, 0)$, $\boldsymbol{v}_d = (v_d, 0, 0)$, and $\boldsymbol{v}_r = \frac{1}{2}(v_d, v_b, 0)$. The net speed and direction – angle θ to the vertical – are given by:

$$|\boldsymbol{v}_r| = \sqrt{\frac{v_d{}^2 + v_b{}^2}{2}} \text{ and } \tan\theta = \frac{v_d}{v_b}$$

[§§] *Dust From Collisions* at *Various Relative Velocities. Akiko M. Nakamura.* Kobe University/CPS at
https://www.astro.uni-jena.de/~theory/DIPS/talks/nakamura.pdf

In the example given above, the resultant speed is about 8 km/sec at an angle of 60^0 to the vertical. Figure 16 shows the layout.

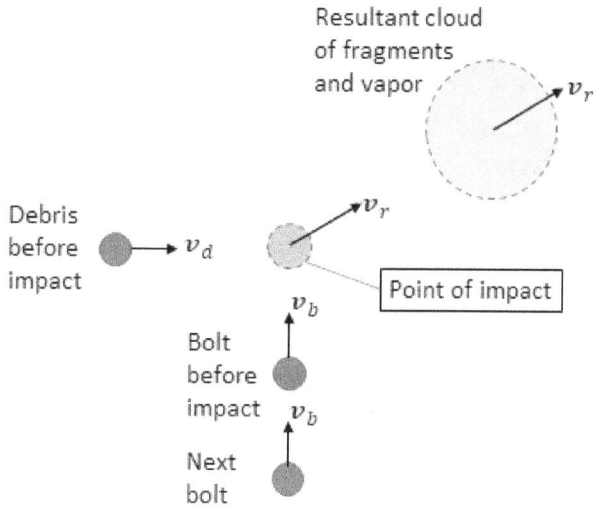

Figure 16 Illustration of an impact, showing before and after

The orbital component of the resultant velocity is $v_d/2$, which is about 5 km/sec in the example. The trailing edge of the cloud of fragments and vapor will travel at about half that speed, while the leading edge will be much faster. Provided the bolt width is substantially less than the spacing between the bolts, the cloud will be out of the way by the time the next bolt arrives.

There will be more complicated cases in which small but significant objects collide with a bolt and may knock them slightly off course. Further study of these cases and a more general treatment is needed. Ultimately, high-speed experiments in a vacuum may be required to verify the theory.

Ascending and descending bolts

Figure 17 shows a view from above a pair of bolts as they pass each other in free space, one descending and the other ascending. In High Stage One, a bolt consists of permanent magnets with electronically controlled electromagnets.

Opposite these are a long chain of permanent magnets designed to keep the bolt in place. In the multi-stage elevator, there are no tubes for most of the way, and the bolts themselves must supply the extra permanent magnets. In the figure, the descending bolt is identical to the ascending bolt but rotated through 180° round the y-axis. Each bolt has two arms, and each arm is similar to a bolt designed for High Stage One. As the bolts pass each other, they are likely to be out of alignment in the x-direction. The permanent magnets are arranged to provide a restoring force to keep them in line. This deals with the Coriolis force, transferring momentum in the x-direction (the orbital direction) from the descending to the ascending bolts. Figures 19 and 20 show a sequence of three time frames as the ascending bolts encounter the descending bolts in free space.

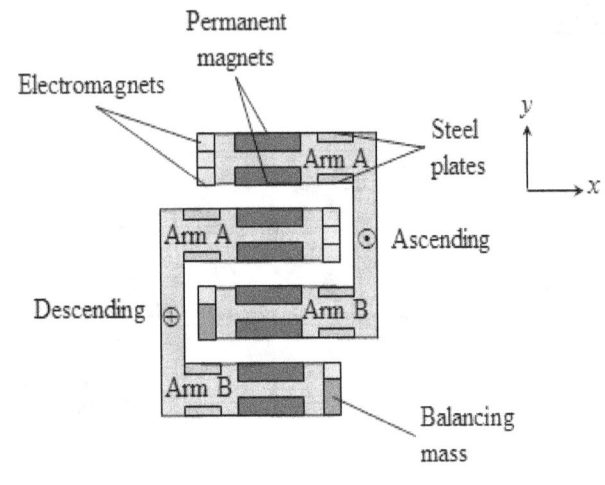

Figure 18 The bolts on the right are ascending; on the left they are descending

International Space Elevator Consortium

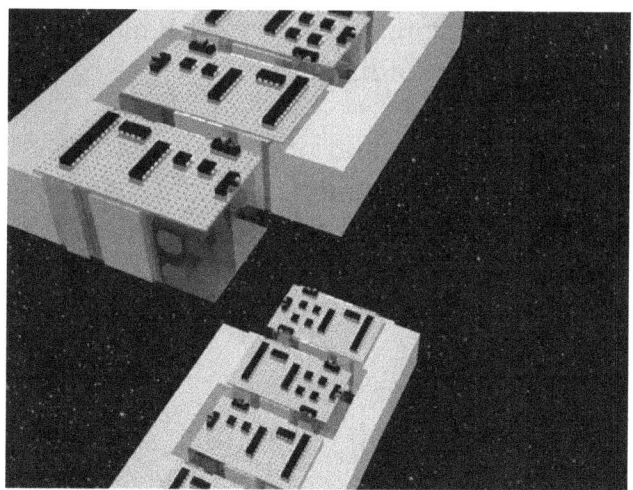

Figure 19 The bolts overlap in this encounter

Figure 20 After the point of overlap, the bolts continue up or down to the next encounter

Permanent magnets facing each other are of opposite polarity to provide the needed attraction in the x-direction. They also attract each other in the y-direction, which is a source of instability. At the right displacement – with a 5 mm gap – they are in balance, but small movements will have to be counteracted by the electromagnets pulling on the steel plates (or similar ferromagnetic material). In theory, the electromagnets should be needed only during the descent. However, they may be required while the bolt is ascending for the sake of rapid response and to dampen any undesired rotational oscillations. There are two electromagnets at the leading edge of Arm A facing in opposite directions. There are two more electromagnets at the trailing edge of Arm A. Arm B contains a single electromagnet at the leading edge and at the trailing edge. The electromagnets in Arm B balance the facing electromagnets in Arm A of the ascending bolt.

There is a "balancing mass" of inert material of the same mass as the electromagnets. This is needed so that the permanent magnets are in line with the center-of-mass of the bolt in order to minimize any tendency to twist or jerk as the bolts approach and pass each other.

5 Earth Port

The Earth port is the point of departure for climbers, and it is the location of the lower ambit. It could be on land but more likely will be at sea. The specialized tube climbers, will ascend the tubes up to the first-stage upper ambit at 100 km altitude, where the payloads will be transferred to tether climbers for the journey to GEO. There are several choices for powering tube climbers, one of which is to draw power from the tubes; another is simply to winch them up from the surface.

The design of the earth port from the 2015 ISEC study is shown in Figure 21. It shows two floating platforms, each of which has a space-elevator tether with a climber on it. A berth for ocean-going ships is shown together with loading and other facilities. The multi-stage space elevator requires these facilities in addition to the large, partially submerged structure called the lower ambit. The other difference is that the tether is replaced by tubes. Climbers ascend the tubes through the atmosphere and transfer to the tether at 100 km altitude.

Figure 21 Floating Tether Terminus Platforms

The other structure that forms part of the earth port is the floating operations platform (FOP), which is the same as described in the 2015 ISEC study. The Tether Terminus Platforms (TTPs) are floating facilities that primarily receive payload transported from the FOP, typically by offshore service vessels*** shown in Figure 22 (taken from the 2015 report) as well as helicopters. Once received, the payloads are loaded aboard the tether climbers and elevated into space.

Figure 22 Offshore service vessel

5.1   Size and configuration

An ambit's job is to turn incoming bolts through 180° and send them on their way. In space, the ambits can use permanent magnets but, on Earth, superconducting magnets are needed. In the atmosphere, the tubes containing ascending and descending bolts must be kept together in order to use active curvature control to maintain stability [5].

Objects of mass m traveling at velocity v round a curve of radius r require a centripetal force $F = m\,v^2/r$ to keep them turning. The magnets in an ambit engage with those in the bolt to provide this force. A free bolt's mass is 0.6 kg. The lower ambit deals with free bolts and tube bolts, but the free bolts are much faster, and so they govern the overall dimensions.

In the lower ambit, the velocity of those bolts intended to rise to 6000 km is 7.87 km/s. An ambit of 6 km radius therefore requires superconducting magnets able to exert a force of 6.2 kN. In a simulation, a force of 13 kN was found, but practical experience is that the forces are usually about half the simulation results, which fits with the 6 km radius for the superconducting part of the ambit. As seen in Figure 23 and Figure 24, this determines the depth to which the ambit has to go below the ocean or the land surface.

*** 140 feet LOA by 29 foot beam by 13 foot draft. Air draft is 93 feet.

International Space Elevator Consortium

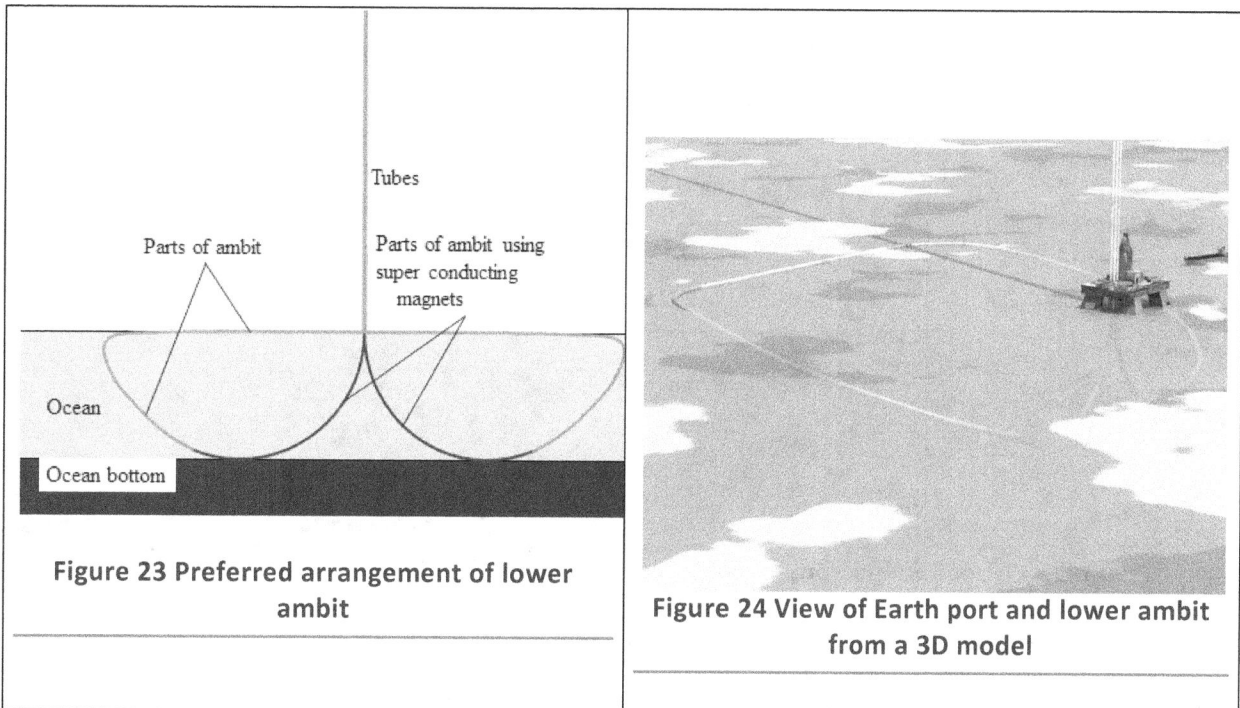

| Figure 23 Preferred arrangement of lower ambit | Figure 24 View of Earth port and lower ambit from a 3D model |

The site proposed in the 2015 ISEC study for the marine node is the island of Kiribati, south of Hawaii. The depth there is a little over 3000 m. A deeper site for the marine node with the lower ambit is an area known as the Gofar Fracture Zone (4° 30' S, 105° 30' W), which plunges to a depth of 6242 m. It is situated about 1600 km south-south-west of the Galapagos Islands. This depth can accommodate the ambit radius of 6 km. Further west, the Central Pacific Basin (9° N, 180° W) reaches a depth of 9047 m, requiring a force in the ambit of only 20 kN.

Figure 23 shows the preferred arrangement of the lower ambit, viewed from the side. Figures 24 and 25 give an overall view of the lower ambit with the point of departure for tube climbers. This layout is designed to minimize the length of tube with the sharpest turns, which will need superconducting magnets with a radius of curvature equal to the ocean depth. These parts of the ambit turn the descending bolts from the vertical to a moderately rising incline; similarly, they turn the bolts at the other end of the ambit from a moderate incline to a vertical ascent. In between, gentler turns can be performed using permanent magnets, which typically require a radius of curvature seven times that achieved by the superconducting magnets.

International Space Elevator Consortium

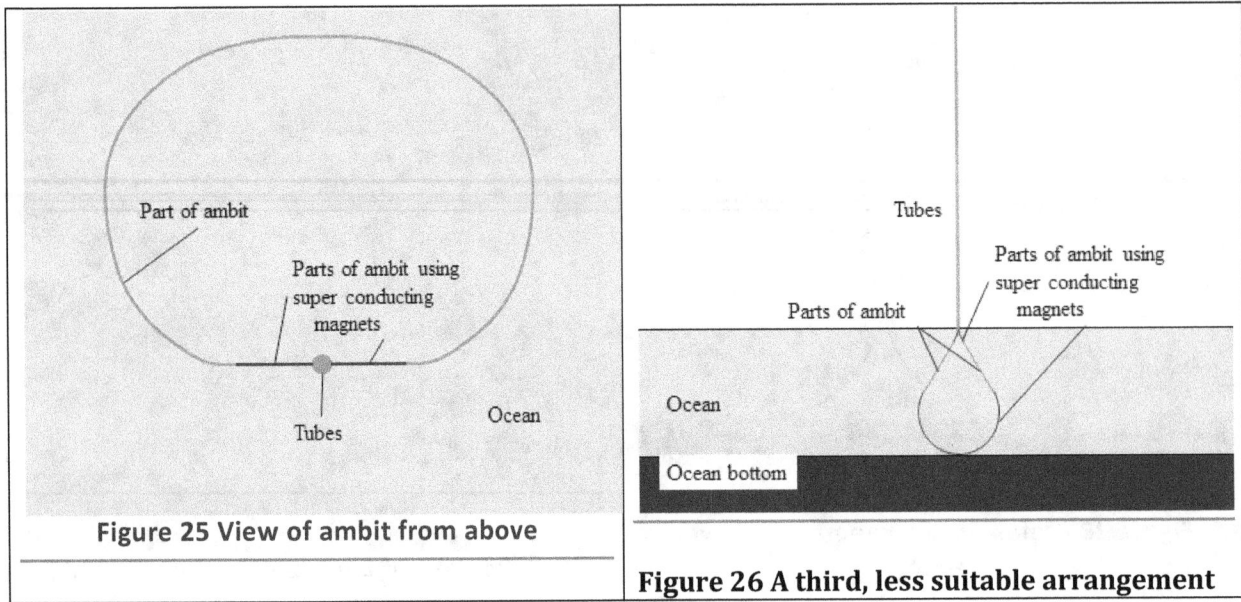

| Figure 25 View of ambit from above | Figure 26 A third, less suitable arrangement |

Figure 25 shows the wide gentle curve of the greater part of the ambit. Using permanent magnets where possible is likely to be less expensive than using superconducting magnets throughout, although a much shorter ambit with tight bends can be achieved using superconductors. This latter option may be preferred, depending on the results of more detailed work.

A third arrangement shown in Figure 26 was rejected, because it would require even tighter curves for a given ocean depth.

5.2   Optimal scaling

As originally proposed, the lower ambit would need to reach a depth of 6000 meters below the ocean surface (or in a deep mine on land), whereas the state of the art in marine engineering is about half that. The original assumptions were that a free bolt would have a mass of 0.6 kg and be 10 cm long. In the lower ambit, the bolt speed is 7870 m/s, which reduces to 1200 m/s at 6000 km altitude. The spacing is inversely proportional to the speed, so that a bolt spacing of 1.3 meters in the lower ambit leads to a spacing of 20 cm in the second-stage ambit. Taking account of the bolt size, that leaves a gap of 10 cm between bolts.

If we halve all three linear dimensions of a bolt, its mass is divided by eight (following the cube law), so that 0.6 kg reduces to 0.075 kg. However, the magnetic attraction is proportional to the surface area, which reduces by a factor of four (following the square law).

Objects of mass m traveling at velocity v round a curve of radius r require a centripetal force $F = m v^2/r$ to keep them turning. Hence, $r = v^2 m/F$ and the ratio m/F reduces linearly with scale. Hence, the radius of the lower ambit can be halved if the linear dimensions of the bolts are halved. This depth of 3000 m allows the earth port with the lower ambit to be situated in the vicinity of Kiribati in the central Pacific.

At the upper ambits, the radius is not an issue. The key requirement is to achieve an upward force strong enough to hold up the tether, the climbers and the other structures that depend on it. That force is proportional to the rate of change of the bolts' momentum, which scales with their mass. However, we can reduce the bolts' spacing in linear proportion to their size reduction, and then the upward force from turning around a stream of bolts will follow a square law, reducing by a factor of four. Therefore, we will need four times as many streams, so that the total number of bolts needed follows a cube law, thus keeping up the same total bolt mass as before.

Table 6 Effects of reducing the scale

Linear scaling	1	1/2	1/4	1/10
Force scaling (square)	1	1/4	1/16	1/100
Mass scaling (cube)	1	1/8	1/64	1/1000
Lower ambit size	6000 m	3000 m	1500 m	600 m
Number of bolts	4 million	32 million	256 million	4 billion

International Space Elevator Consortium

6 Design of the upper stages

The most exacting requirements come from the lower ambit, which will probably be at sea. Bolts arrive and depart vertically and must be turned around as rapidly as possible in order to minimize the depth of the overall structure. This requires superconducting magnets in the region of greatest curvature. Once the arriving bolts are turned through approximately 120^0, permanent magnets are sufficient to guide them to near the surface and turn them around.

In the upper stage ambits, however, permanent magnets are the only realistic option because of the high degree of maintenance required for superconducting magnets. Because neodymium magnets are three times as powerful as ceramic ferrite magnets, their use makes the upper ambits much lighter. One way of using neodymium magnets is to make them of many narrow rods each coated with a thin insulating layer. Then eddy currents have almost nowhere to go. This method is suggested in academic literature but does not appear to have been adopted by manufacturers. It would require a custom production run, which would be worthwhile for the larger and full-size versions of the multi-stage space elevator but not justified for the small prototypes.

It is possible to use ceramic ferrite magnets in the tubes and bolts and use neodymium in the ambits. This gives a tripling of the force between them compared to using ferrites throughout. Using neodymium throughout gives a factor of nine increase in force.

The main purpose of an ambit is to turn the incoming bolts around so that they return in the direction from which they came. This is to be achieved with minimal loss. Turning the ascending bolts around so that they descend creates a strong upward force which supports the structures.

6.1   First stage just above the atmosphere

The first stage ambit has to support its own weight and the tubes beneath it, as well as whatever installations are required on the first-stage platform for transferring payloads to the tether climbers. It may be that a tube climber is simply a box containing the tether climber, or the two types of climber may differ, depending on the chosen mechanism for raising payloads in the atmosphere. It has to support the mechanism that raises tube climbers from the surface, which could be a winch, a specialized tether or a linear induction system, possibly drawing power from the kinetic energy of the bolts traveling inside the tubes. It also has to support the weight of the Whipple or similar shielding against debris and radiation above the atmosphere.

International Space Elevator Consortium

Bolts arrive at the first stage ambit in evacuated tubes from the Earth's surface. Each tube contains two straight parallel magnetic tracks, which are mainly designed to stop the bolts colliding with the sides of the tube. They are also part of the system for maintaining stability in the presence of gusting winds in the atmosphere. These magnetic tracks match the permanent magnets and electromagnets in the tube bolts, but they do not need much strength. A single line of 5×10 mm ferrite magnets will serve well.

The ambits contain continuations of the magnetic tracks, but they need much greater strength. When a bolt arrives at the ambit, it encounters curved tracks. It follows the curve, turning with it rather like a train turning with the tracks. The magnets are arranged in Halbach arrays matching the Halbach arrays in the bolts. Bolts of mass m traveling at velocity v round a curve of radius r require a centripetal force $F = m v^2/r$ to keep them turning. Hence, $r = v^2 m/F$. Assume a force F of 400 N and a mass m of 0.4 kg for a tube bolt. If $v = 1.4 \times 10^3$ m/s, r is about 2 km. The length of track in the ambit is given by $2\pi r$, about 13 km.

To calculate the total upward force on the ambit due to the passing bolts, we need to consider the rate of change of momentum of the stream of bolts. Turning one bolt around gives a change of momentum of $2mv$. If n_a bolts arrive at the ambit and n_a bolts leave the ambit every second, the rate of change of momentum of the stream of bolts at the ambit is $2n_a mv$. This is the upward force exerted by turning the bolts around. If there are n bolts per meter traveling at velocity v, then nv bolts are turned around by the ambit every second. Therefore, $n_a = nv$ and the rate of change of momentum is $2nmv^2$.

If the bolts are spaced 0.2 m (20 cm) apart, there are five bolts per meter and $n = 5$. Then the upward force on the ambit is $F_a = 2 \times 5 \times 0.4 \times (1.4 \times 10^3)^2 \cong 8 \times 10^6$ N, i.e., 8 MN, equivalent to just over 800 tonnes weight. A section of prototype ambit weighs 7 kg per meter, and so 13 km of ambit would weigh about $13 \times 7 \cong 90$ tonnes. Add another 60 tonnes for structural elements to leave 650 tonnes force available to support tubes, vehicles and other objects. This is the force from a stream of bolts ascending and descending.

In the atmosphere, the evacuated tubes will need to be fairly heavy to withstand weather while maintaining the vacuum, so it is preferable to have multiple streams of bolts in a single evacuated tube. This provides a nice balance if we arrange for four ascending and four descending streams in one tube. A tube mass of 10 kg per meter leads to a tube mass of 1000 tonnes over its 100 km length. To this must be added the Whipple shield extending down for 20 km from the first stage. This gives an additional 10 kg per meter mass, amounting to an extra 200 tonnes. With this design, we have four streams of bolts exerting a combined upward force of 2600 tonnes weight supporting 1200 tonnes of tube and shielding, leaving 1400 tonnes (about 14 MN) of available force to support other structures and vehicles. The other structures include the tubes for the bolts that will proceed to the upper stage or stages as well as shielding for the first stage.

If 1400 tonnes is not enough, a second set of four ascending and four descending streams of bolts may be added to give another 1400 tonnes of support capacity.

6.2   Second stage or higher

The discussion deals with the two-stage design. Similar considerations apply to more stages. The second stage does not have to support tubes, and its own weight per meter is much less than that of the first stage owing to the diminished gravity at 6000 km. The bolts will leave the lower ambit at 7.87 km/s, but they will arrive at the second stage at about 1.2 km/s spaced at 5 bolts per meter. By a calculation similar to that for the first stage, they will create an upward force of 5.7×10^6N. As in the first stage, the ambit mass will still be 150 tonnes, but this weighs only 390 kN, leaving 5.3 MN for lifting. The force required to support the tether below 6000 km with climbers at 100 km, 1580 km and 4600 km (the worst case scenario with climbers limited to a constant power of 4 MW) is 9 MN, so two ascending and two descending streams of free bolts are sufficient.

6.3   Entering or leaving the ambit

Figure 27 shows a section of the second-stage ambit, which has tracks that match the magnets on the free bolts. The curve of the ambit can be seen in this picture, and the bolts follow this curve as they travel and so make the required turn. The ascending bolts arrive, and the descending bolts depart.

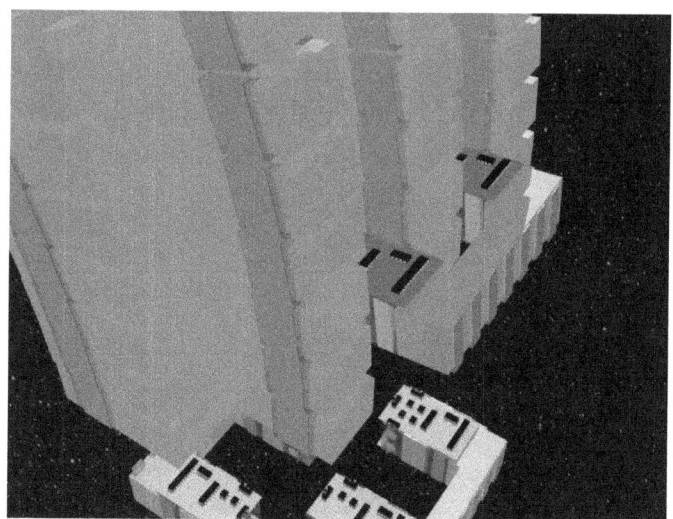

Figure 27 A section of the second-stage ambit

Figure 28 free bolts traveling in two opposite directions in space

However, in free space, the ascending and descending bolts are magnetically engaged as in Figure 28. At the ambit, the arriving bolts must be pulled away from the departing bolts to make room for the ambit's magnetic tracks to be inserted between them. The arrangement is illustrated in Figure 29.

A similar, but inverted, arrangement is needed at the point on the first stage where the bolts destined for the higher stage leave the evacuated tubes and enter free space. At that point, they must engage with the bolts that are descending from the higher stage.

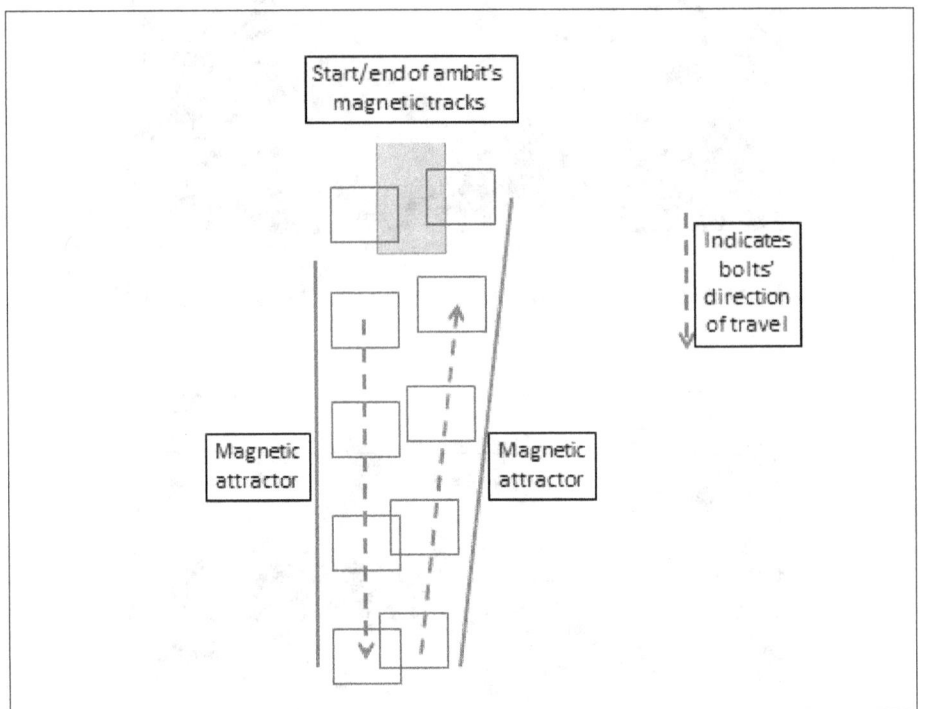

Figure 29 Arrangement at the ambit entrance

6.4   Alternative ambit design

The shape of the upper ambits has been shown as in Figure 30. The streams of bolts travel close to each other, whether in tubes in the atmosphere or in free space. An alternative design which is simpler, lighter and smaller is shown in Figure 31. There, streams of bolts, both ascending and descending, are in two separate groups, while the tether is suspended in between.

International Space Elevator Consortium

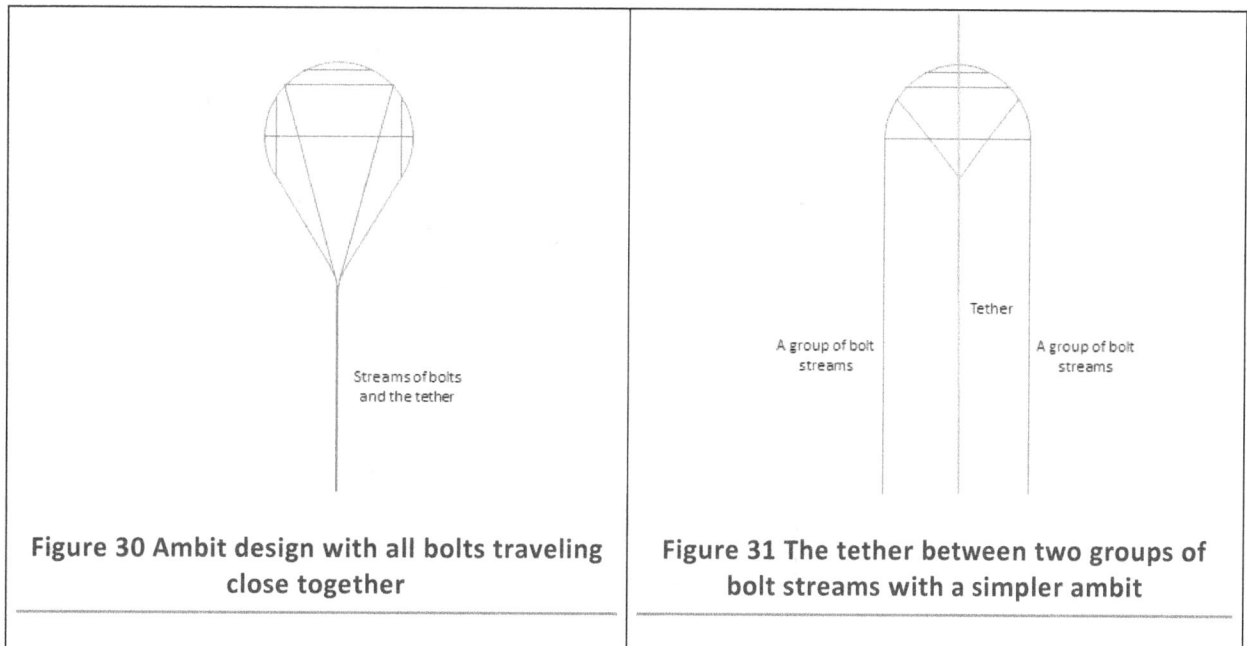

| Figure 30 Ambit design with all bolts traveling close together | Figure 31 The tether between two groups of bolt streams with a simpler ambit |

6.5   Construction methods

The plan is to raise two pairs of tubes together with their upper ambit to 100 km altitude. The first pair of tubes and ambit are raised to 200 m height. They then support the second pair of tubes, which are raised to 400 m height. The pattern then is to alternate, doubling the height each time, until the desired altitude of 100 km is reached. So the intermediate steps are 200, 400 and 800 m; 1.6, 3.2, 6.4, 12.8, 25, and 50 km.

To erect the second pair of tubes and ambit, we haul them up to the height of the first ambit and then accelerate the bolts inside them until they reach the necessary speed to support the ambit at the higher altitude. Then we allow them to rise to double the altitude of the first pair, which takes place in increments of 100 m.

Once a pair of tubes and an ambit have reached 100 km altitude, they can support the erection of further tubes, each in a single step, to establish all the infrastructure needed for the first stage. This includes the tubes needed for the second stage and higher. Once the first stage infrastructure is established, the higher stages can be erected in a similar series of alternating steps, doubling the altitude each time until the required altitudes are reached. We raise each part of the tether with the ambit that supports it.

International Space Elevator Consortium

Alternating steps in the atmosphere

The reason for alternating and doubling the height at each step is to ensure that enough bolts are present to support the ambit. The spacing between bolts is about 10 times their length at the earth's surface but shrinks as the bolts slow down during their ascent. As we lengthen the tubes, the spacing between the bolts increases. Accelerating the bolts mitigates this effect until a point is reached at which there would no longer be enough bolts to hold up the ambit. That point occurs at some time after the tube lengths have doubled.

When the first pair of tubes has reached its altitude of 200 m, the next step is to use it to raise the second pair of tubes and ambit with double the number of bolts inside. Then we accelerate them to the desired speed to cause them to rise to double its height, i.e., to 400 m. The first pair of tubes now has to be lowered and extended to the length of 400 m with enough bolts to reach 800 m. Then we haul them up to 400 m and accelerate the bolts to the speed needed to reach 800 m. The process continues up to 100 km altitude.

Raising the first pair of tubes

The first-stage ambit supports the weight of a pair of tubes due to the large upward force created as it turns the bolts inside it from traveling up to traveling down. To raise the tubes, we have to increase the speed of the bolts. We can hold the tubes under increased tension while we accelerate the bolts until they are traveling fast enough to exert the upward force required. Then we allow the ambit and the pair of tubes to rise under this increased force.

Figure 32 shows an early stage of construction. The lower ambit below the surface of the sea (or underground in a deep mine) has to be built first. The ambits and tubes are next evacuated, while the upper ambit is supported by scaffolding or similar temporary structures. The thrusters in the lower ambit then accelerate the bolts traveling inside it to the required speed to support the upper ambit. Then the construction support can be removed. Next the vacuum vessel is evacuated, and the upper ambit rises due to the bolts traveling inside it. Figure 33 shows the next step, where the temporary magnetic tracks hold the bolts in line inside the evacuated chamber.

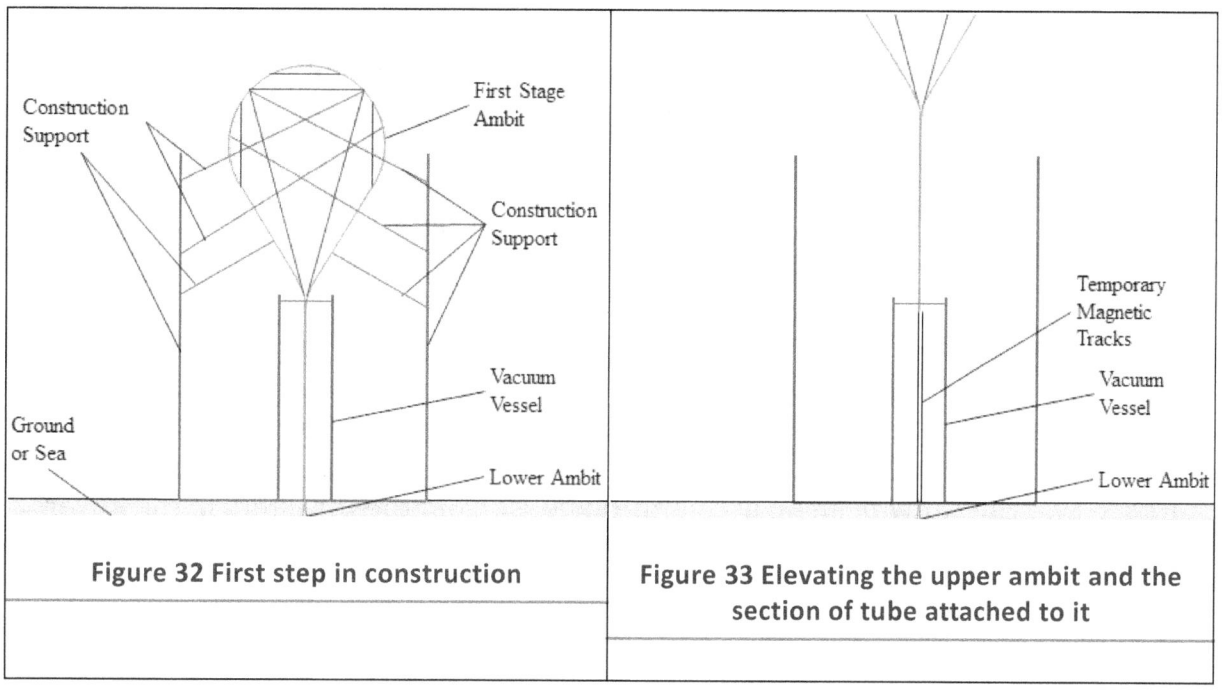

Figure 32 First step in construction

Figure 33 Elevating the upper ambit and the section of tube attached to it

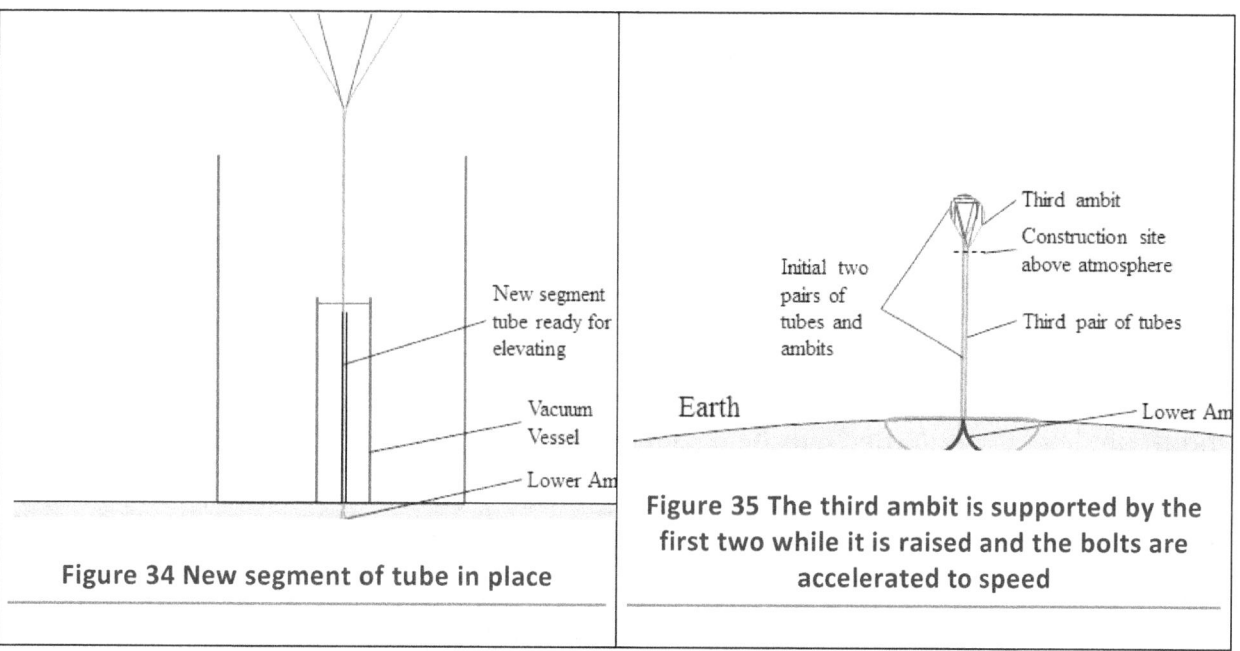

Figure 34 New segment of tube in place

Figure 35 The third ambit is supported by the first two while it is raised and the bolts are accelerated to speed

International Space Elevator Consortium

Once the upper ambit and tube have been elevated to the height shown, a new segment of tube is spliced to the existing tube inside the vacuum vessel (Figure 34). The new segment may be composed of several shorter lengths, each of which is split longitudinally into two halves which are brought together inside the vacuum vessel. The height of the vacuum vessel is proposed as 100 meters. The seals at the top of the vacuum vessel must retain their integrity throughout this process.

Raising the second pair of tubes

We haul the second ambit and tubes up to the height of the first. Then we have to accelerate the bolts. It is possible to do this from the surface, but it may be preferable to haul the bolts up inside the tubes and then release them from the ambit so that they only need an acceleration boost from the thrusters in the lower ambit. Otherwise, they have to be accelerated from a standing start in the thrusters to a great enough speed to reach the upper ambit.

When the second ambit has reached the height of the first, it can be raised in 100 m stages using the same technique as for the first tube until it reaches double that height.

Then the first tube is lowered and extended to double its length with four times the number of bolts, and the process is repeated until the altitude of 100 km has been reached.

Another method

Instead of alternating between two ambits, each with their pair of tubes, another approach is to add to the number of bolts in a tube by means of sophisticated electronic controls in the thruster. The requirement would be to insert a new bolt at full speed midway between every two existing bolts as they pass through. This involves high precision timing at high speed, but it has the advantage that it can be perfected in a laboratory before construction begins. From an overall engineering point of view, it would perhaps be easier than raising alternating ambits.

Raising the third pair of tubes

Once the first two pairs of tubes have been erected so that their upper ambit is above the earth's atmosphere at 100 km, we can use them to raise the third ambit and pair of tubes without needing the vacuum vessel on the earth's surface. The first two ambits and pairs of tubes can support the third (Figure 35) until the bolts inside them are accelerated to the velocity required to support them.

We carry on erecting ambits and tubes as necessary and then put in place the tubes and other infrastructure needed to build and manage the upper stages.

Erecting the upper stages

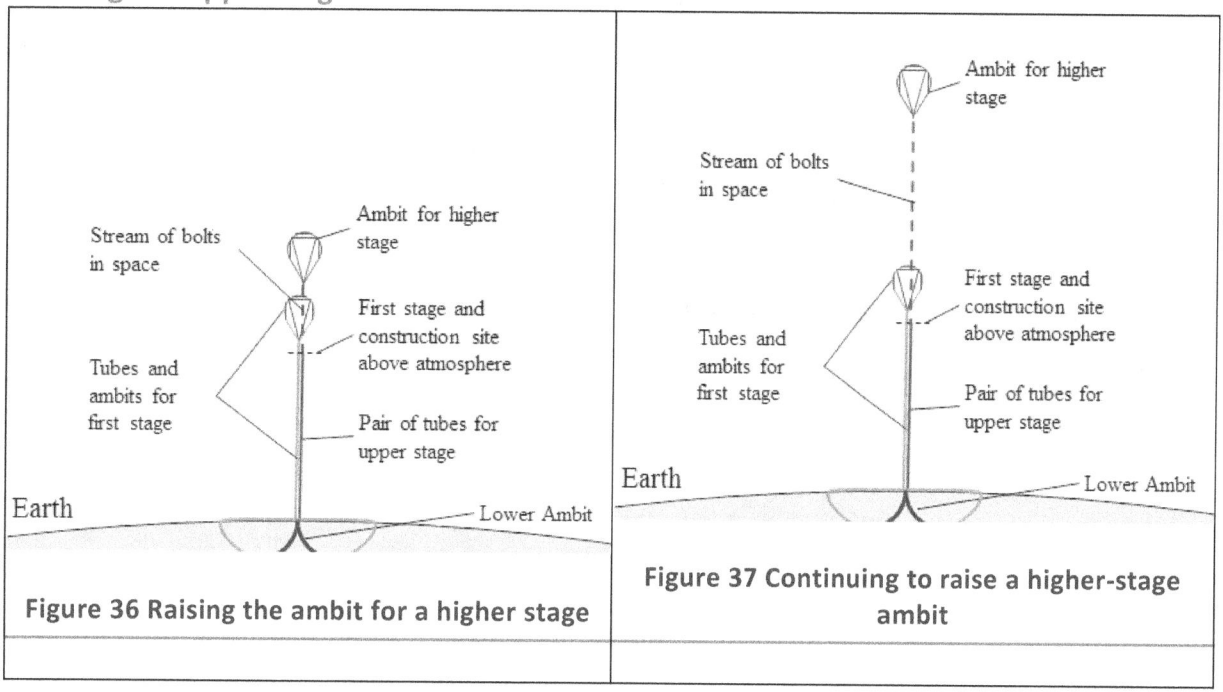

Figure 36 Raising the ambit for a higher stage

Figure 37 Continuing to raise a higher-stage ambit

The second (or higher) stage needs the first stage to support its tubes up to 100 km. Above that, the bolts travel in the vacuum of space without tubes. Once the first stage is in place, we raise the second-stage ambit on to it and align it with the appropriate pair of tubes. Then, the lower ambit (below the surface) accelerates bolts in the tubes until they are fast enough to support and then raise the ambit that is destined for the second or higher stage (Figure 36).

The process continues (Figure 37), gradually accelerating the bolts until the upper ambit is in the right place for the second or higher stage. For robustness and resilience, at least two pairs of tubes are needed for each stage.

Keeping it stable during the erection process

When complete, the tower can be held stable by the tether reaching down from the Apex Anchor (which acts as a counterweight), but during erection it will need thrusters at the upper ambits to maintain stability.

Connecting to the tether

Above the highest ambit, the tether is supported by the balance of forces up to the Apex Anchor.

Once the tubes are in place, we have to lower a slim tether from GEO, as described in the Edwards book [10]. This tether will require thrusters to move it to the point where it connects with the tether supported by the bolts and ambits. Then small construction climbers can ascend it to supply additional material to augment the tether until it reaches full strength.

Accelerating a bolt

During erection, it will be necessary to accelerate the bolts at the surface station so that they can reach the altitude of the first step and return to Earth. If the first step is 100 m high inside an evacuated chamber, the bolts in step 1 only need a modest speed of 50 m/s initially. After that, they can be accelerated gradually as the ambit rises. To give a bolt a linear acceleration, a good method is to use linear motors as thrusters.

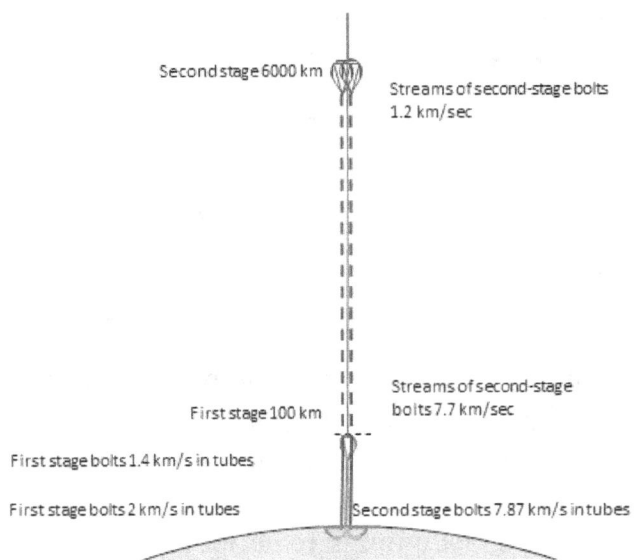

Figure 38 Two stages with bolt speeds

To reach 100 km, the bolts need a speed of 2 km/s (Figure 38). Accelerating them to this speed from a standing start is technically possible, but it may be simpler to haul a set of bolts up to 100 km altitude and let them fall under gravity, accelerating as they go. At the bottom, their kinetic energy will be close to that needed to send them back up, and the surface station has an easier task to give them the necessary boost to reach the top again, rather than boosting their energy from zero at the surface. During normal operation the bolts will need a speed boost at the surface station – and possibly at the upper ambit – to make up for friction or other losses.

7 Development program with cost estimates for prototypes

The approach to costing has been first to seek commercial estimates for building the main components and then to estimate the additional work and facilities needed to assemble and test the prototype.

Earlier proposals to build prototypes of High Stage One have been adapted for the multi-stage space elevator. As for High Stage One, several steps are proposed, beginning with a bench-top or bench-side version and progressing to an indoor structure 10 meters high. The next step is an outdoor structure 60 meters high.

7.1 List of Tasks with GANTT chart
The following tasks appear in the GANTT chart in Figure 39.

Theoretical work
1. Examine collision and cascade scenarios in space, particularly those near to an entry into a tube or ambit. The aim is risk assessment and mitigation.
2. Assess the consequences of air leaks leading to loss of vacuum, and study how to minimize the risk.
3. Improve the work on stability in the atmosphere, adapting the more advanced technique published at the 2017 SEC, which measures the displacement between bolts traveling in opposite directions and is included in the 2018 ISEC study report. The 2009 BIS publication described a method called active curvature control that relied on measuring the bending of a tube and the wind force. The intent of the improved method is to rely solely on measuring the displacement between a bolt and a tube.

Prototyping
1. Build a non-working model for exhibition and presentation purposes.
2. Prepare and find sources of supply for electromagnets, 3D printing, circuit board layout, electronic circuit assembly, IR (or other) sensors, batteries, and power capacitors.

3. Construct a working bench-side prototype.
    a. Startup and shutdown procedures
    b. Design of bolt trains
    c. Bolt manufacture
        i. Hand assembly from components
        ii. Testing
    d. Optimizing IC code
    e. Design, build, test and placement of thrusters
    f. Track construction
4. Design, build and test a bolt that can recharge its batteries and capacitors in flight using an induction coil.

Miniaturization
Choose the best approach to miniaturizing the bolts.
1. Select the target scale reduction (between 2:1 and 10:1)
2. Design the miniature bolts
3. Find a supplier and validate the manufacturing capability

Evacuated Tubes
This phase builds on the experience and skills accumulated from the first prototype. Continuity of at least some personnel is highly desirable.
1. Design the version that operates in evacuated tubes.
    a. Seals and pumps
    b. Tube material and construction
    c. Selection of materials in bolts and thrusters to avoid outgassing
2. Construct a version up to 10 meters high in an engineering lab, a warehouse or a similar building.

Vacuum Experiments
1. Design, refine, build and test the free bolts, i.e., the M-shaped bolts intended to operate in free space.
    a. Spinning wheel test in a vacuum chamber with bolts on two wheels spinning in opposite directions, increasing the wheels' speed as much as possible.

b. Test of two streams of bolts traveling in free space in opposite directions. The speed will be limited by the size of the vacuum chamber.

2. Assess the effectiveness of the stability method in a vacuum chamber.
3. Experiment with collision scenarios in the vacuum chamber.
4. Devise and execute tests in orbit, including verifying the method of dealing with the Coriolis force.

Costing of Development

1. Produce cost estimates for the development phases. Cost estimates are available for the first two phases, bench-side (2 meters) and evacuated tubes (10 meters). We need estimates for evacuated-tube structures at successive heights of 60 meters, 1 km and 10 km.
2. Cost the vacuum chamber tests
3. Estimate the cost of constructing a full-size ambit
4. Produce cost estimates for the final process of construction and erection.
   a. Building the tubes up to 100 km
   b. Creation of first-stage platform at 100 km
   c. Elevation of the second stage to 6000 km
   d. Additional work involved in building up to five stages to accommodate progressively weaker tether materials

International Space Elevator Consortium

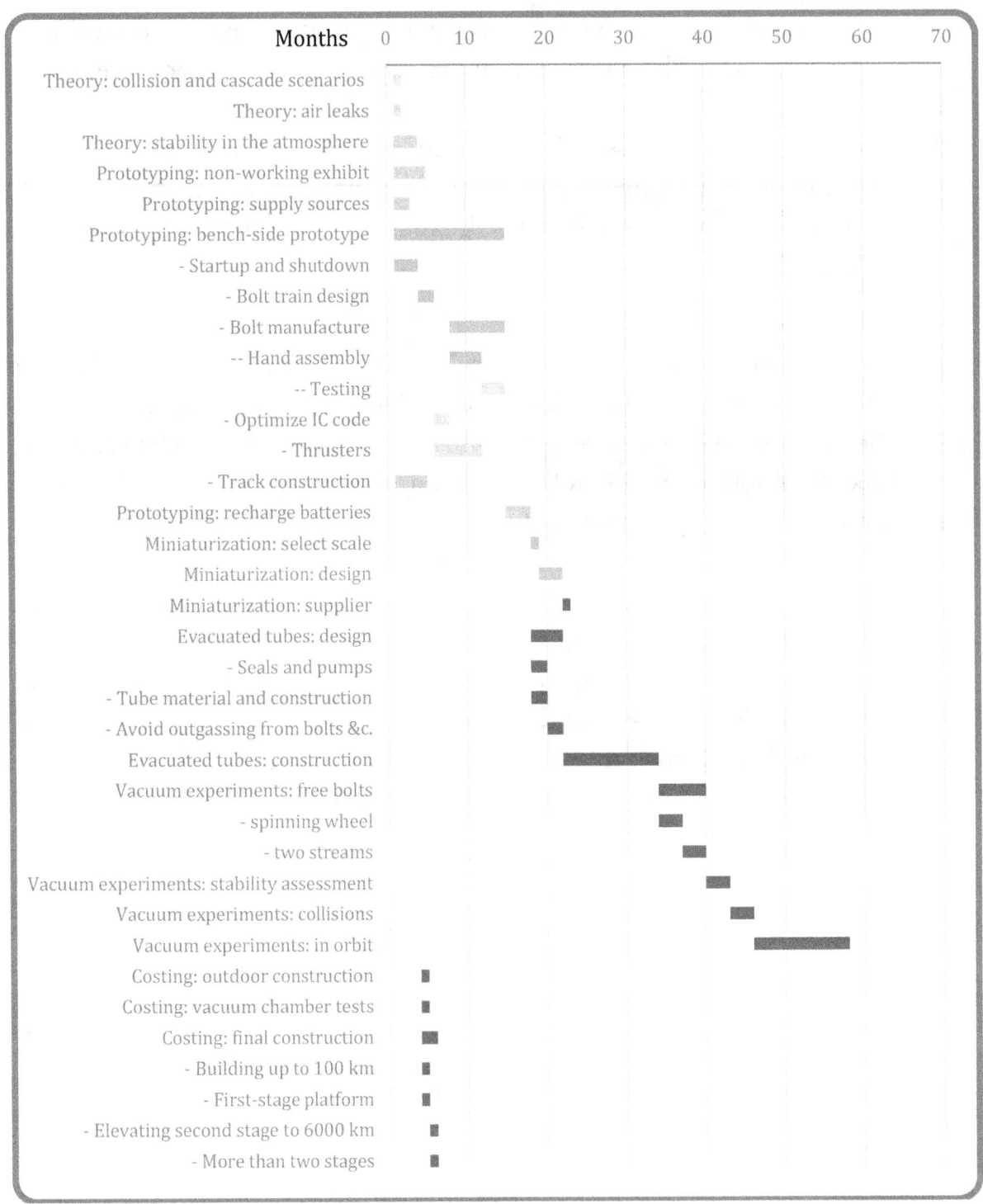

Figure 39 GANTT chart of the tasks

7.2   Bench-side version

This is intended to be a proof of concept for dynamically supported structures. The prototype (Figure 40) reaches a height of 4 meters. The height of the initial structure is 3 meters, and it must be supported by a frame. The bolts inside will initially be at rest and then will be accelerated. When the required velocity is reached, the supporting frame will be removed and the upper ambit will rise. In the process, the length of the middle section between the ambits will double. This is achieved by allowing the elements of the middle section to move apart using a simple sliding mechanism.

All this is performed in air. The aim of the experiment is to assess vibration, component reliability and the published method of erecting the structure. The length of track that the bolts pass along in a single circuit before the upper ambit rises is 9 m. The bolt speed is 9.5 m/s, reducing to 7 m/s at the top. Each bolt is 10 cm long, and there are five bolts per meter at the bottom, increasing to 6.8 per meter at the top. There are about 55 bolts in all.

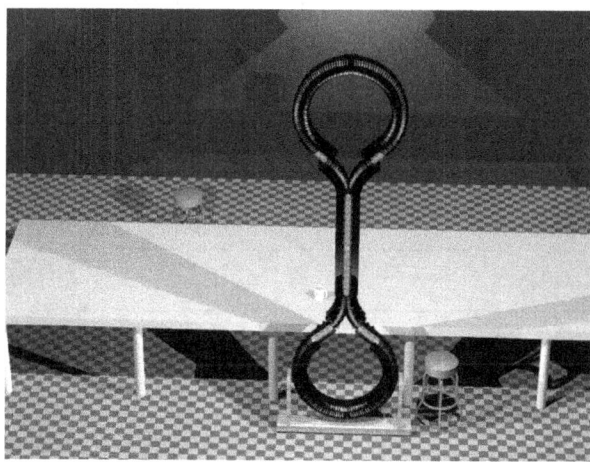

Figure 40 First prototype of the multi-stage space elevator

7.3   Costs

The component costs are as follows: 160 electromagnets are estimated to cost $2000 (£500+£500+ about £450); they are needed both for the bolts (55 bolts, 2 per bolt) and the thrusters (50). 3×55 Li-Ion batteries (£14 each) cost $3500.

Each bolt requires 6 ferrite permanent magnets 40×10×10 mm – a total of 6×55=330. The 9 m track requires 6 of the same magnet per 40 mm of track, giving a total of 1350 magnets. Overall, just under 1700 magnets are needed. The estimate from IMA to supply the magnets is €1500 or $1800. Bunting Magnetics Europe have given an estimate of $7750 for 600 of the assembled arrays of three magnets each.

Each bolt will need two different PCBs with the relevant components on each. Each circuit's components cost about $25, and we need 2×55. To this we add 2×50 thruster circuits to make a total of 210 circuits at $25, amounting to $5250. The cost of the PCBs must be added – three different designs – both for fabrication and assembly. The quote from PCBWay (www.pcbway.com) in China is approximately $1400 including shipping. A similar service from PCBTrain (www.pcbtrain.co.uk) in the UK costs about $6000.

Item	Year 1 Cost US$
Electromagnets	2,000
Batteries	3,500
Magnet arrays	7,750
PCBs and assembled circuits	7,400
Bulk 3D printing	40,000
Subtotal components	60,650
Tools	50,000
Staff member at 2×$100,000 per annum	200,000
Lab/workshop fees	50,000
TOTAL	**360,650**

Table 7 Costs for first prototype

Bulk 3D printing of the structural components is the largest outsourced activity, coming in at $40,000. A trade-off can be made between outsourcing and employing labor in house to operate the 3D printers. Other tools will need to be purchased, such as two or three 3D printers and electronic and mechanical tools, for which we budget $50,000.

Finally, we need to consider employing someone to assemble and test the complete prototype. This is the big item and will require the use of lab or workshop facilities for up to a year, which must be added to the cost. We estimate $200,000 per person per year, plus lab fees of $50,000.

These costs are summarized in Table 7.

7.4   10-meter version using evacuated tubes

The next version uses evacuated tubes and is illustrated in Figure 41.

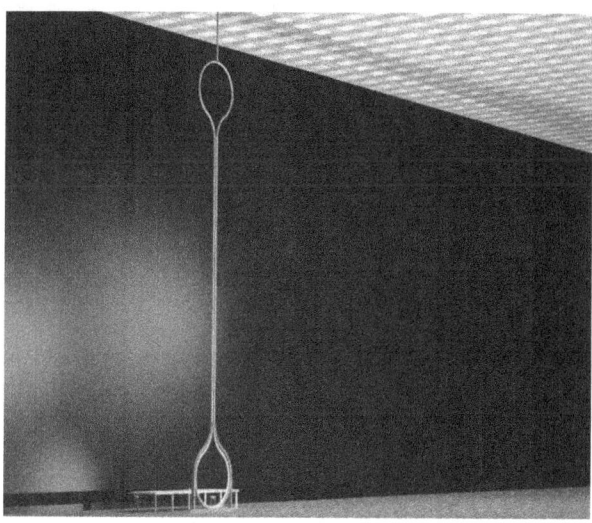

Figure 41 Prototype 10 meters high

The component costs are roughly three times those for the first prototype, plus the cost of vacuum equipment and tubes. Two years are required with two staff members, plus the help of a vacuum consultant for the first phase of the work. The consultant's costs will be about $60,000, and the vacuum equipment will be about $50,000. The tubes will cost another $20,000. These considerations lead to the estimates in Table 8.

Item	Year 2 Cost US$	Year 3 Cost US$
Electromagnets	6,000	0
Batteries	10,500	0
Magnet arrays	23,250	0
PCBs and assembled circuits	22,200	0
Bulk 3D printing	120,000	0
Vacuum tubes incl. allowance for replacement	20,000	20,000
Subtotal components	281,950	20,000
Tools	100,000	10,000
Vacuum equipment	50,000	10,000
Vacuum consultant	60,000	10,000
Two staff members at 2×$100,000 each per annum	400,000	400,000
Lab/workshop fees	100,000	100,000
TOTAL	**991,950**	**550,000**

Table 8 Costs of 10-meter prototype

7.5   Overall costs

The annual costs are summarized in Table 9.

Table 9 Annual development costs

Year 1	Year 2	Year 3	Total
$360,650	$991,950	$550,000	$1,902,600

The total cost over three years comes to $1.9 million.

7.6   Costs of development stages

If we take the cost of the 10-meter version to be $1.55 million, we can obtain a first-approximation estimate by considering the construction of successive versions each 10 times as big as the last one and 10 times the cost, as in Table 10.

Version size	Cost estimate (US$ millions)
10 meters	1.55
100 meters	15.5
1 km	155
10 km	1550
100 km	15500
Total	**16,222**

Table 10 Cost estimate of increasing sizes up to 100 km

To this figure of $16 billion must be added the cost of the streams of bolts that rise to 6000 km, which will cost about $2.8 billion. Finally, we need the first and second stages. The ambits are costed at $1.5 billion each, and the ancillary equipment needed at the first stage will cost another $800 million. This brings the total to $22.6 billion.

8 Conclusions and recommendations

If we had access to the very strong material envisaged by Edwards and others, we would not need the complexity of multiple stages with dynamic support. However, the present work shows that a stable multi-stage structure is possible and could potentially be engineered to use small low-cost components.

The two-stage architecture using an 11 MYuri material is obviously preferred over a higher number of stages, but the choice is there to have more stages and use progressively weaker materials. Much depends on whether we can build and test more prototypes of bolts and other components. If we invest in the multi-stage approach as well as in stronger materials, we will increase the likelihood of being able to build a space elevator sooner rather than later.

References

[1] Yunxiang Bai, Rufan Zhang, Xuan Ye, Zhenxing Zhu, Huanhuan Xie, Boyuan Shen, Dali Cai, Bofei Liu, Chenxi Zhang, Zhao Jia, Shenli Zhang, Xide Li and Fei Wei, "Carbon nanotube bundles with tensile strength over 80 GPa", *Nature Nanotechnology* (2018).

[2] Forward, R., *Indistinguishable from Magic*, Baen Publishing Enterprises, Riverdale, NY, 1995, pp. 59-89.

[3] Lofstrom, K., *The Launch Loop*, AIAA Paper 85-1368, July 1985.

[4] Knapman, J. and Swan, P., "Design Concepts for the First 40 km – A Key Step for the Space Elevator," *Acta Astronautica* (2013), DOI: 10.1016/j.actaastro. 2014.06.004

[5] Knapman, J., "The Space Cable: Capability and Stability," *Journal of the British Interplanetary Society*, Vol. 62, No. 6, 2009, pp. 202-210

[6] Knapman, J., "Stability of the Space Cable", *Acta Astronautica* 65 (2009) pp. 123–130.

[7] Swan, P. A., Raitt, D. I., Swan, C. W., Penny, R. E. and Knapman, J. M., *Space Elevators: An Assessment of the Technological Feasibility and the Way Forward*, International Academy of Astronautics, Paris, 2013.

[8] Rufan Zhang, Qian Wen, Weizhong Qian, Dang Sheng Su, Qiang Zhang and Fei Wei, "Superstrong Ultralong Carbon Nanotubes for Mechanical Energy Storage," *Advanced Materials* 2011, *23*, 3387–3391

International Space Elevator Consortium

[9] Wei Xu, Yun Chen, Hang Zhan, and Jian Nong Wang, "High-Strength Carbon Nanotube Film from Improving Alignment and Densification," *Nano Lett.* 2016, 16, pp. 946–952.

[10] Edwards, B. and Westling, E. (2003), *The Space Elevator,* BC Edwards, Houston, Tx.

[11] Ishikawa, (2013) The Space Elevator Construction Concept (Obayashi Corporation), Yoji Ishikawa, Tatsuhito Tamura, Kiyotoshi Otsuka, Takaya Horiike, Takeo Iwaoka, Naoki Masui, Katsuya Hamachi, Satomi Katsuyama and Yoshio Aoki, IAC-13-D4.3.6

[12] Shelef, B., "The Space Elevator Feasibility Condition," *Climb*, Vol.1, No.1, p.87, The International Space Elevator Consortium (2011).

[13] Kristian Smistrup, Peter T. Tang, Ole Hansen and Mikkel F. Hansen, "Microelectromagnet for magnetic manipulation in lab-on-a-chip systems," *Journal of Magnetism and Magnetic Materials* 300, pp. 418–426 (2006).

[14] Knapman, J., "Diverse Configurations of the Space Cable", *61st International Astronautical Congress,* Prague, Czech Republic, September 27 – October1, 2010.

[15] J. Knapman, "Stability of the Space Cable", *57th International Astronautical Congress*, Valencia, Spain, October 2006.

Appendix A   International Space Elevator Consortium

Who We Are
The International Space Elevator Consortium (ISEC) is composed of individuals and organizations from around the world who share a vision of humanity in space.

Our Vision
A world with inexpensive, safe, routine, and efficient access to space for the benefit of all mankind.

Our Mission
The ISEC promotes the development, construction and operation of a space elevator infrastructure as a revolutionary and efficient way to space for all humanity.

What We Do
- Provide technical leadership promoting development, construction, and operation of space elevator infrastructures.
- Become the "go to" organization for all things space elevator.
- Energize and stimulate the public and the space community to support a space elevator for low cost access to space.
- Stimulate science, technology, engineering, and mathematics (STEM) educational activities while supporting educational gatherings, meetings, workshops, classes, and other similar events to carry out this mission.

A Brief History of ISEC

The idea for an organization like ISEC had been discussed for years, but it wasn't until the Space Elevator Conference in Redmond, Washington, in July of 2008, that things became serious. Interest and enthusiasm for a space elevator had reached an all-time peak and, with Space Elevator conferences upcoming in both Europe and Japan, it was felt that this was the time to formalize an international organization. An initial set of directors and officers were elected and they immediately began the difficult task of unifying the disparate efforts of space elevator supporters worldwide.

ISEC's first Strategic Plan was adopted in January of 2010 and it is now the driving force behind ISEC's efforts. This Strategic Plan calls for adopting a yearly theme to focus ISEC activities. (For 2010, the theme was "Space Elevator Survivability -- Space Debris Mitigation.") In 2010, ISEC also announced the first annual Artsutanov and Pearson prizes to be awarded for "exceptional papers that advance our understanding of the Space Elevator." Because of our common goals and hopes for the future of mankind off--planet, ISEC became an Affiliate of the National Space Society in August of 2013.

Our Approach

ISEC's activities are pushing the concept of space elevators forward. These cross all disciplines and encourage people from around the world to participate. The following activities are being accomplished in parallel:

- Yearly conference – International space elevator conferences were initiated by Dr. Brad Edwards in the Seattle area in 2002. Follow--on conferences were in Santa Fe (2003), Washington DC (2004), Albuquerque (2005/6 –smaller sessions), and Seattle (2008 to the present). Each of these conferences had multiple discussions across the whole arena of space elevators with remarkable concepts and presentations. Recent conferences have been sponsored by Microsoft, the Seattle Museum of Flight, the Space Elevator Blog, the Leeward Space Foundation, and ISEC.
- Yearlong technical studies – ISEC sponsors research into a focused topic each year to ensure progress in a discipline within the space elevator project. The first such study was conducted in 2010 to evaluate the threat of space debris. The second study, and resulting report, focused on space elevator operations. The 2013 study focused upon tether climber designs. The 2014 topic is Space Elevator Architectures and Roadmaps. There is one topic chosen for 2015; Earth Port Design Considerations. The products from these studies are reports that are published to document progress in the development of space elevators. They can be downloaded at www.isec.org.
- International cooperation – ISEC supports many activities around the globe to ensure that space elevators keep progressing towards a developmental program. International activities include coordinating with the two other major societies focusing on space elevators: the Japanese Space Elevator Association and EuroSpaceward. In addition, ISEC supports symposia and presentations at the International Academy of Astronautics and the International Astronautical Federation Congress each year.
- Publications – ISEC publishes a monthly e--Newsletter, its yearly study reports and an annual technical journal (CLIMB) to help spread information about space elevators. In addition, there is a magazine filled with space elevator literature called Via Ad Astra.
- Reference material – ISEC is building a Space Elevator Library, including a reference database of Space Elevator related papers and publications.

- CLIMB – This annual peer reviewed journal invites and evaluates papers and presents them in an annual publication with the purpose of explaining technical advances to the public. The first issue of CLIMB was dedicated to Mr. Yuri Artsutanov (a co-inventor of the space elevator concept); and, the second issue was dedicated to Mr. Jerome Pearson (another co--inventor). CLIMB is scheduled for publication each July. They can be downloaded at www.isec.org.
- Outreach – People need to be made aware of the idea of a space elevator. Our outreach activity is responsible for providing the blueprint to reach societal, governmental, educational, and media institutions and expose them to the benefits of space elevators. ISEC members are readily available to speak at conferences and other public events in support of the space elevator. In addition to our monthly e--Newsletter, we are also on Facebook, Linked In, and Twitter.
- Legal – The space elevator is going to break new legal ground. Existing space treaties may need to be amended. New treaties may be needed. International cooperation must be sought. Insurability will be a requirement. Legal activities encompass the legal environment of a space elevator -- international maritime, air, and space law. Also, there will be interest within intellectual property, liability, and commerce law. Starting work on the legal foundation well in advance will result in a more rational product.
- History Committee – ISEC supports a small group of volunteers to document the history of space elevators. The committee's purpose is to provide insight into the progress being achieved currently and over the last century.
- Research Committee – ISEC is gathering the insight of researchers from around the world with respect to the future of space elevators. As scientific papers, reports and books are published, the research committee is pulling together this relative progress to assist academia and industry to progress towards an operational space elevator infrastructure. For more visit http://isec.org/index.php/about-isec/isec-research-committee
- Competitions – ISEC has a history of actively supporting competitions that push technologies in the area of space elevators. The initial activities were centered on NASA's Centennial Challenges called "Elevator: 2010." Inside this were two specific challenges: Tether Challenge and Beam Power Challenge. The highlight came when Laser Motive won $900,000 in 2009, as they reached one kilometer in altitude racing other teams up a tether suspended from a helicopter. There were also multiple competitions where different strengths of materials were tested going for a NASA prize – with no winners. In addition, ISEC supports the educational efforts of various organizations, such as the LEGO space elevator climb competition at our Seattle conference. Competitions have also been conducted in both Japan and Europe.

ISEC is a traditional not-for-profit 501 (c) (3) organization with a board of directors and four officers: President, Vice President, Treasurer, and Secretary. In addition, ISEC is closely associated with the conference preparation team and other volunteer members. Address: ISEC, PMB 204, 9272 Jeronimo Rd Ste 107A, Irvine, Ca 92618-1978 inbox@isec.org / www.isec.org

Appendix B   Acronyms and Lexicon

CNT	Carbon Nano Tube
FOC	Full Operational Capability
FOP	Floating Operations Platform
GEO	Geosynchronous Earth Orbit
HQ/POC	Headquarters Primary Operations Center
IAA	International Academy of Astronautics
IOC	Initial Operational Capability
ISEC	International Space Elevator Consortium
kg	kilogram
LOA	Length Overall
MT	metric ton
MW	megawatt
NASA	National Aeronautics and Space Administration

IAA study group #3-24 met in Seattle in August of 2015. The team agreed to use, as much as possible, consistent terminology for this report. Below are those terms shown in the figure. . This general list of terminology is shown in the next table: The agreed upon terms should be:

Apex Anchor Node	LEO Gate	Earth Port
Mars Gate	Lunar Gravity Center	- Earth Terminus
Moon Gate	Mars Gravity Center	- Floating Operations Platform
GEO Node	Tether Climbers	Headquarters and Primary Operations Center

Figure - Space Elevator System Lexicon Example

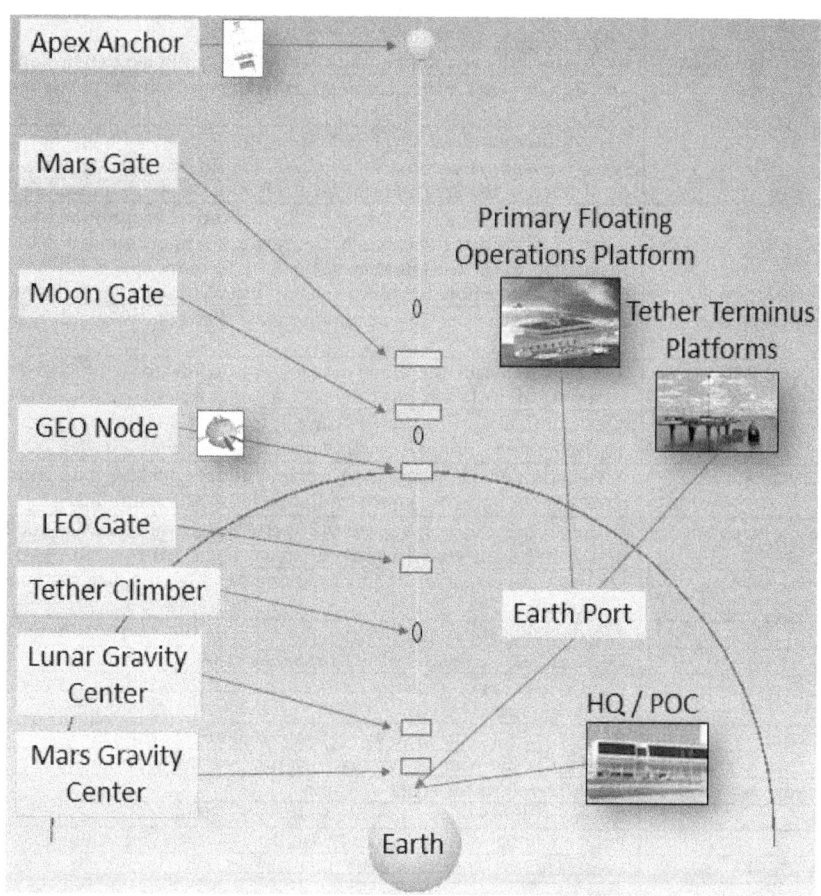

Table, Lexicon of Terms

International Space Elevator Consortium

Terminology	Explanation
Apex Anchor	The upper end at roughly 100,000 kms altitude providing counterweight stability for the space elevator as a large end mass, with elements such as; reel in/out capability, thrusters to maintain stability, command and control elements, etc..
Apex Anchor Region	The region is the volume swept out by the end of the tether during normal operations. When two or more space elevators are operating together, the region spreads to the volume between.
Boron-Nitride Nanotube (BNNT)	High Tensile Strength material under development
Capability On Ramps leading to FOC	Time after IOC when new businesses / capabilities are added to system [7th sequence step]
Carbon Nanotube (CNT)	High Tensile Strength material under development
Climbers [Tether Climbers]	Vehicle able to climb or lower itself on the tether
Deployment	Releasing the tether from the GEO construction up and or down during the initial phase of construction
Earth Anchor	Earth Terminus for space elevator
Earth Port	Consists of Earth Anchor [terminus] and Floating Operations Platform and has a complex required to support its functions.
Earth Port Region	The volumetric region around each Earth Port to include a space elevator column for each tether and the space between multiple tethers when they operate together. The Earth Port Region will include the vertical volume through the atmosphere up to where the space elevator tether climbers start operations in the vacuum and down to the ocean floor.
Final (Full) Operational Capability	Design for full capability of the space elevator [8th sequence step]
Floating Operations Platform	The Op's Center for the activities at the Marine Node or Earth Terminus [Earth Port]
GEO Node	Geosynchronous Earth Orbit (GEO) Facility – roughly 36,000 kms altitude – for space elevator systems control and customer support. The complex of activities positioned in the Space Elevator GEO Region; directly above the Earth Port.
GEO Region	Encompasses all volume swept out by the tether around the Geosynchronous altitude, as well as the orbits of the various support and service spacecraft "assigned" to the GEO Region. When two or more space elevators are operating together, the region includes each and the volume between elevators.
Headquarters and Primary Operations Center [HQ/POC]	Location for the Operations and Business Centers – probably other than at Marine Node
Initial Operational Capability	A term to describe the time when the space elevator is prepared to operate for commercial profit – robotically [6th sequence step]
International Academy of Astronautics (IAA)	International Association focusing upon space capabilities with approximately 1,000 elected members.
International Space Elevator Consortium (ISEC)	Association whose vision is: A world with inexpensive, safe, routine, and efficient access to space for the benefit of all mankind.
Japanese Space Elevator Association	JSEA handles all the space elevator activities for universities and STEM activities. Also handles the global aspects of space elevators.
Japanese Space Agency (JAXA)	Japanese government organization responsible for space systems and space operations.
Length Overall	Full length of the space elevator, est. from 96,000 to 100,000 km
LEO Gate	Elliptical release point for LEO – roughly 24,000 kms altitude
Limited Operational Capability	Early utilization of a "starter" tether in parallel with testing and further development [5th sequence step]
Lunar Gate (Moon Gate)	Release Point towards Moon – roughly 47,000 kms altitude
Lunar Gravity Center	Point on Tether with Lunar gravity similarity – 8,900 kms altitude
Marine Node (Earth Port)	Earth Terminus for space elevator
Mars Gate	Release Point to Mars – roughly 57,000 kms altitude
Mars Gravity Center	Point on Tether with Mars gravity similarity – 3,900 kms altitude
Ocean Going Vehicle (OGV)	Vehicle able to travel over the open ocean
Operational Testing	Key developmental phase when checking out capability [4th sequence step]
Pathfinder	In-orbit testing of space elevator with as many segments represented as possible [1st sequence step]

International Space Elevator Consortium

Primary Operations Center	Center of all activities for the space elevator. Could be distributed or centralized.
Seed Tether [Ribbon]	The initial tether lowered from GEO altitude which would then be built up to become the space elevator tether [2nd sequence step]
Single String Testing	Single string tests are tests conducted of a selected set of Space Elevator functions; aligned and operating. In early forms, single string testing could be an end-to-end simulation of a segment. Later, hardware is inserted in the string to add realism. Testing the initial tether after deployment would be a key single string test.
Space Elevator Column	The volume swept out during normal operations starting at the Earth Port [a circular area within which it operates] and extending through the GEO Region up to the Apex Region. This column of space will be monitored, restricted, and coordinated with all who wish to transverse the volume.
Tether	100,000 km long woven ribbon of space elevator with sufficient strength to weight ratio to enable an elevator [CNT material probably]
Tether Climbers	Vehicle able to climb or lower itself on the tether, as well as releasing or capturing satellites for transportation or orbital insertion.

Terminology specific to the multi-stage space elevator

Figure - Layout of major components

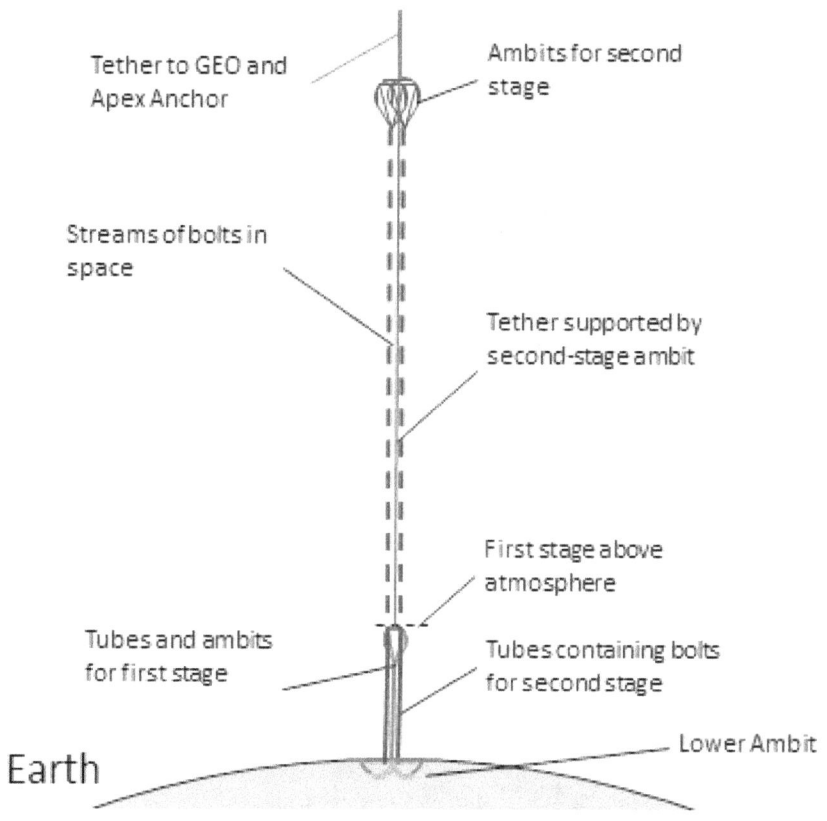

Tether to GEO and Apex Anchor

Ambits for second stage

Streams of bolts in space

Tether supported by second-stage ambit

First stage above atmosphere

Tubes and ambits for first stage

Tubes containing bolts for second stage

Lower Ambit

Earth

International Space Elevator Consortium

Tether climber	An electric vehicle that carries a payload from the first stage to geosynchronous altitude.
Bolt	A small object that travels at high speed. Turning it round from ascending to descending causes an upward force that supports the structures, the tether and the tether climbers.
Tube	Evacuated tubes in the atmosphere provide a low-friction environment for the bolts. Magnetic levitation prevents the bolts colliding with the sides of the tubes.
First stage	A structure at 100 km altitude, which is the edge of space, supported by bolts traveling at high speed in evacuated tubes. It supports the tubes below it.
Upper stage	A structure in space thousands of kilometers from the earth's surface that supports the tether below it. It is supported by bolts traveling at high speed in the vacuum of space.
Second stage	The lowest of the upper stages. Depending on the tether material used, it is between 1500 and 6000 km from the earth's surface.
Higher stage	A stage above the first stage
Tube climber	An electric vehicle that carries a payload from the earth's surface to the first stage. It may carry a complete tether climber or a container that is ready to be attached to a tether climber at the first stage.
Ambit	A horseshoe-shaped structure that turns bolts around using magnetic levitation.
Lower ambit	The lower ambit is evacuated and is submerged at sea. Descending bolts enter it from the tubes, turn around, receive a speed and power boost, and start to ascend in tubes.
Upper ambit	Each upper stage has an upper ambit that receives ascending bolts and turns them around so that they descend. The force thus created holds up the tether. The upper ambits are in space and so are naturally evacuated.
High Stage One	An earlier proposal similar in shape to the Launch Loop. It relieves the tether of the unpredictable forces in the atmosphere. The first stage of the multi-stage space elevator supersedes it.

International Space Elevator Consortium

Appendix C Architecture Engineering Baseline Change Management

"Now that we know what we want to do ... how can we change it?"
Michael A. Fitzgerald

Introduction

The Year 2017 was a big year. It was so big that it has taken nearly 3/4 of 2018 to get our story straight; if we could just get people to join with us and our vision, all would be good. Right?

Not really, we need to change or improve a few things and **then** all would be good. Right? Not really, we will need to change or improve a few more things and then all would be good.   Right? Well ... maybe; but I doubt it. Therefore, I have an announcement.

We should be prepared for a decade of changes, followed by another series of changes, and after that, more changes. So we need an orderly approach to dealing with change. An immutable change approach is essential so that everyone working on the Space Elevator is working on the same thing. The "**change approach**" matures into a detailed **change process** to be used during the space elevator's design and development.

It is a little early to get involved in documenting the change process now, but we must be aware that it is coming. Our modular design approach (See Architecture Note #1) and our application of sequenced Architecture Engineering principals (See Architecture Notes #6 through #8) must adhere to the fundamental Architecture theme that mission performance in one segment can affect mission performance in other segments. Thus, we foresee a **change approach** which is used between now and the end of validation demonstrations (see the Architectures and Roadmap Report), and a **change process** used during design and development.

So, What!?

We are going to use the two terms to differentiate change during the current time frame and then change during later time frames. The "**change approach**" is operative now and helps us monitor what our baseline is for the Space Elevator Transportation System. For the most relevant example, the Earth Port of the current Space Elevator Transportation System baseline presumes a tether material sufficiently strong to allow direct connection to the Earth Port's tether termini.

The current investigation of the Multi Stage Space Elevator by John Knapman's team is not the baseline approach. However, John's investigation may well portend that it is a better baseline. In the same way, the investigation of the graphene sheeting approach may prove a viable tether solution for our Space Elevator Transportation System. We shall see.

If evidence appears that one or the other portends functional success, the ISEC Chief Architect will declare a **"Call for Improvement"**. In that Call, and under our **change process**; we will seek to assess which approach serves best as our baseline. A series of validation experiments and demonstrations will be the basis of ISEC's assessment, showing that the needed performance can be attained.

In simple terms, what would be next?

Should the Multi Stage Space Elevator prototype project show evidence that it offers the best path to IOC, then its technology would be seen as verified. Following that, a new baseline Earth Port would be defined to replace the one in the current baseline. The same process would be followed for graphene sheeting.

In closing

Change is good. The current baseline will be defined under another title.

Fitzer

Appendix D   Brainstorming Session Minutes

Here are the questions for the mini-workshop held on August 17th, 2018 at the Space Elevator Conference in Seattle:

- What prototype phases should we plan?
    - How big does a prototype need to be to be convincing?
- What should the funding balance be between strong materials and multi-stage technology?
    - Suppose we had $1 million, $10 million or $100 million.
- What are good methods of descent?
    - Falling, gliding, retro rockets?
    - Coming down the tether?
    - What about jumping or crossing over ascending climbers?
- Propose good operating procedures
    - Use automation and remote control as much as possible

Following are the results.

Prototyping
1. Component prototypes for each bit, particularly bolt and ambit
2. Magnetic friction to be tested in a continuous ring (5m, 50m) in vacuum at high speeds
3. Destructive testing to understand where things break
    a. When things hit the bolts from the side (deflections)
    b. When the bolt hits the ambit (misses the entrance)
4. The acceleration phase, getting the bolts up to speed. Build an accelerator to inject bolts into the ring. Suggested a horizontal, linear accelerator
5. Prototypes do not necessarily need to be complete, but representative.
6. Testing the bend in the plane of accelerated bolts from horizontal to vertical to send up to the ambits
7. How to scale the smaller prototypes to the larger in providing the currents for the magnetic fields. At some point you need to prototype superconducting technology to accelerate the bolts. There will be a jump in how to provide the increase in energy.
8. Each prototype needs to be simulated and then backed up with lab data
9. Define what it is we are prototyping
10. Prototype the deployment approach (adding the tubes to raise the upper ambit).

11. Build an amusement park ride based on the bolt/ambit concept
12. Demonstrate how long the system runs without power.
13. What happens when the vacuum fails in the tubes (another failure mode to test)
14. Demonstrate climber transfer between stages
15. Maintaining vacuum at depth (6km?) in the sea
16. How big should a prototype be to convince people?
    a. Not too big that press can't see the top
    b. Big enough to demo superconducting magnets to imply scale from there
    c. Some people may never be convinced, even after you build the real thing

Contributors: Dennis Wright, Adrian Nixon, Phil Swan, Sean Sun, Scott Snowden, David Horn, Michael Fitzgerald

Funding

<u>Overall and General Group Thoughts:</u>
1. Strong materials - if you got the materials development close to completion, then you wouldn't need to do a multi-stage.
2. Do you spend the money on research or on final completion?
3. Find the criteria for how much we should put towards A or B? Society can always use materials for other applications.
4. It would be great if we could have only one stage - that way you can build it quicker out of weaker materials.
5. Modular self-assembling systems - something that could build itself from the ground up is really interesting.
6. One of the problems with the modular systems - is it can fall if it's ever turned off.
7. Most of the money should go towards materials development - because the stronger the material gets, the lower you can bring the tether down towards the earth.
   - Do you build it in house or out of house? Spread the interest around the world.
8. How do we spend the money? If you start doing something today or later with stronger materials?
9. The first ambit will not launch anything into space and is ready to support the thing above it.
10. Can it be made strong enough to turn it into a LEO launch platform?
11. Additional commercial value to getting to space - tourism, research, enterprise, solar, shipyard

12. Specifically aimed at holding up the tether below it.
13. But if you're waiting for that then you're vastly extending the timeline.

<u>Individual Scenarios:</u>
What are the criteria of how to spend the money between Materials (A) and Multi-stage (B)?

1. What is best for humans: how good it is for humans as a species? - 90% into Materials, 10% into multistage. There is so much more that we can intrinsically benefit from having strong materials within the context of the space elevator. 10% into multi-stage development because investors never put all their money in one bucket - diversification of risk. Space elevator becomes the carrot at the end of the stick.
2. If we have a viable Lofstrom loop already - Lowers the risk of Multi-stage considerably, the technology risk is gone. You're adapting existing technology to a new application. 75% into Multistage, and 25% into Materials development because then it can help lower the overall cost down the timeline faster. Demonstrated technology.
3. If we have to get to space now - Alien attacks, asteroid impact, etc. Assuming not rockets: Tough question. The high was 80% multi-stage and 20% materials development since we have existing technologies now that might be able to get us there. But that went all the way to 50/50 because there are other ideas that can be combined to make the multistage concept a reality. Another thought: build rotating tethers approach.
4. Which would attract the most money down the road?

Contributors: Ruth, Nigel, David, Bryan, David, Drew

Methods of Descent
The discussion reached across many options and discussed the range of alternatives that seem to cover the waterfront. Here is a list of discussion topics with short explanations:

1. Option – Only upward. The idea is that all tether climbers only go up and the design incorporates "reuse" of all hardware at GEO and beyond. This enables the tether to NOT DEAL WITH the tremendous energy buildup in heat of climbers coming down the tether and using brakes. The concept recommends that all "valuable cargo" that needs to come back to the Earth (people and high value components) with rockets – of course the rockets had been delivered to GEO by the tether so the cost is low.
2. Tether Braking: The concern here is that coming down the tether creates heat as the brakes are applied. The tremendous amount of energy in the tether climber [potential and kinetic energies] must be dissipated as the

climber brakes... usually in heat. The complexity is the amount of heat needed to be dissipated, the impact of the brakes on the tether material (repair, update of tether needed as a result), and the speed resulting from the gravitational pull. Major studies must be accomplished once the material is identified with questions like friction coefficient? Effectiveness of braking? Damage to material? Speed? One aspect of downward is you could save energy in storage such as flywheels or large batteries.

3. Drop off along the way: One concept is to use braking for the low gravity region (say down to one radius high – 6378 km altitude) and then release from tether and reentry into the atmosphere. This would imply that the tether climber is designed differently going down vs going up. The lower portion (high gravity region) would require slowing down with ablative material (?) and then parachutes to land, or aerospace shape for landing and slowing down.

4. Use the Heat (energy): there are many needs for energy, so take the inherent energy from entering the energy well and transfer to users. Some of the potential users are energy projection back to GEO operations or down to surface of Earth or to another satellite or tether climber to energy system. In addition, there is a suggestion to radiate energy at the proper wavelength for growth of ozone for helping our environment.

5. Leverage Multi-Stage Space Elevator: The problem of energy dissipation is largely solved if we have a structure up from the ground to 6,000 km or 15,000 km. Braking on the space elevator seems reasonable if we do not go down to the heavy gravity region (possibility defined as one Earth radii altitude).

6. Rotating Space Elevator: If the baseline design of the space elevator is a two-strand tether that rotates then braking is not an approach. You attach to the rotating tether, go up with the tether, release at your destination, then downward direction payloads attach to the tether and ride down to the release location. The bottom line is: no braking required. This is still a viable approach, just has not been accepted as the "baseline" for a few years.

7. Thrusters slowing down tether climbers: to slow down inside the high gravity well has historically been accomplished by rockets. The suggestion is that downward space elevator tether climbers have assistance to slow down – namely rocket fuel and rocket thrust. We can have 'cheap" fuel by getting it from the Moon or other space resource and delivery to Apex Anchor or GEO Region. Then the thruster is used when the speed becomes too large when going down. This lowers the stress on the tether through braking; however, the thrust vector must NOT be in the direction of the tether – maybe up to 45 degrees out from tether (less efficient, but safer for the tether).

8. Once in the atmosphere: When we are going rapidly inside the atmosphere, there are methods of reducing velocity with parachutes, large drag area structures, or even wings and landing capability.
9. Multi-Leg: One alternative in the design of the space elevator is to have multi-base leg architecture (legs coming together at 2,000 km ???) As one goes up, there is a principle travel leg while the others provide stability and safety/backup. For returning payloads, the high gravity well suggests "slow trips" over the last region. So a different leg coming down would allow the tether climber to proceed slowly and lower its impact on health of the tether material. Slow works better with respect to heating and braking... so come down on another leg – maybe two weeks from 2,000 km.

Operating Procedures

Ongoing Operation
1. Intermodal transfer - automated transfer of payloads from the various process climbers. Surface to S1 climbers and from S1 climbers to S2 climbers, etc.
2. Climbers will remain within their designated process stages
3. Hand off should be automated. Cargo is moved from one vehicle suitable for atmospheric conditions to one suitable for outer space.
4. Handoff from upper stages is merely transfer from tether to tether if it has been staggered.
5. Humans will likely need to reside at each stage for stage maintenance.
6. If the cargo container is pressurized, a human could ride with the vehicle or else wear a pressurized suit.
7. Maintenance: damage is automatically detected and there is an automated repair process. (The process is out of scope.)

Appendix E Mathematical Support

Stability is a key issue for dynamically supported structures. Unless carefully managed, it is easy for a small deviation from the required trajectory to expand indefinitely. Fortunately, there are remedies that are not difficult to implement. The mathematical analysis of stability is presented here in detail.

Mathematics of stability in the atmosphere

In the atmosphere, the bolts travel in tubes and are stabilized using a technique called *active curvature control*, which was developed for an earlier proposal called the Space Cable. It is described in detail in the published literature [5].

Mathematics of stability in space

Permanent magnets provide the main force between bolts traveling in opposite directions. It is usually attractive, but it can be repulsive. However, to maintain stability, some damping is required – mostly quite small but nevertheless essential to avoid oscillations that would otherwise become excessive. The bolts flow in streams, and we find that their flow is governed by fourth-order partial differential equations strongly influenced by the force between bolts traveling in opposite directions.

First, we derive the partial differential equations and then find methods of constraining the lateral motion. We build on previous work carried out for the Space Cable and High Stage One using Laplace transforms. The equations there are second order, but fourth-order equations govern the motion in the multi-stage elevator. The following is the main equation. Its derivation and solution follow below.

$$V \frac{\partial^2 R_b}{\partial x \partial t} + m \left(\frac{\partial^4 z_s}{\partial t^4} - 2V^2 \frac{\partial^4 z_s}{\partial x^2 \partial t^2} + V^4 \frac{\partial^4 z_s}{\partial x^4} \right) = 0 \qquad (1)$$

The displacement along the direction of travel is x, time is t, and the vertical velocity is V. This equation governs the motion of two streams of bolts traveling in opposite directions with a force R_b between them. The displacement z_s is the mean displacement of the pair of streams, i.e., the mid-point between the streams. R_b is the average force per unit distance between the streams. Each stream has a mass m per meter.

The distance between the streams is z_b. For stable damped harmonic motion in the gap, we need a restoring force of $A z_b$, a damping force $c \frac{\partial z_b}{\partial t}$, and some additional terms that we will designate Y.

$$R_b = m \left(Y - A z_b - c \frac{\partial z_b}{\partial t} \right) \qquad (2)$$

A and *c* are constants, as are l, g and h in the following definition of Y:

$$Y = m\left(\frac{l}{V}\int \frac{\partial z_s}{\partial t}dx + \frac{g}{V}\int z_s dx - 2cV\frac{\partial z_s}{\partial x} + \frac{h}{V}\iint z_s dxdt - 2AV\int \frac{\partial z_s}{\partial x}dt\right) \qquad (3)$$

We have assumed that m and V are slowly varying compared to z_s and z_b. The same assumption applies to the external force R_e, which is due mainly to tides caused by the sun, moon and planets.

$$R_e = 2m\left(\frac{\partial^2 z_s}{\partial t^2} + V\frac{\partial^2 z_b}{\partial x\partial t} + V^2\frac{\partial^2 z_s}{\partial x^2}\right) \qquad (4)$$

The solution is stable provided certain conditions on A, c, l, g and h are met.

Simulation

A simulation of an 8 km length of a pair of streams is shown. In this simulation, a disturbance is applied simultaneously and in the same direction both at the top and the bottom. Snapshots show the movement of the upward (blue) and downward (orange) streams as the effect is transmitted by the bolts traveling at 1.7 km/sec. The yellow box with a blue edge shows the simulated time. The vertical scale is in km; the horizontal is in mm.

In Figure 42, both the top and the bottom are deflected, and the movement is propagated. An algorithm detects and corrects for the deflection at the top and bottom. The point of the simulation is to ensure that the effect is attenuated, or at least not amplified, as the movement travels from top to bottom or vice versa.

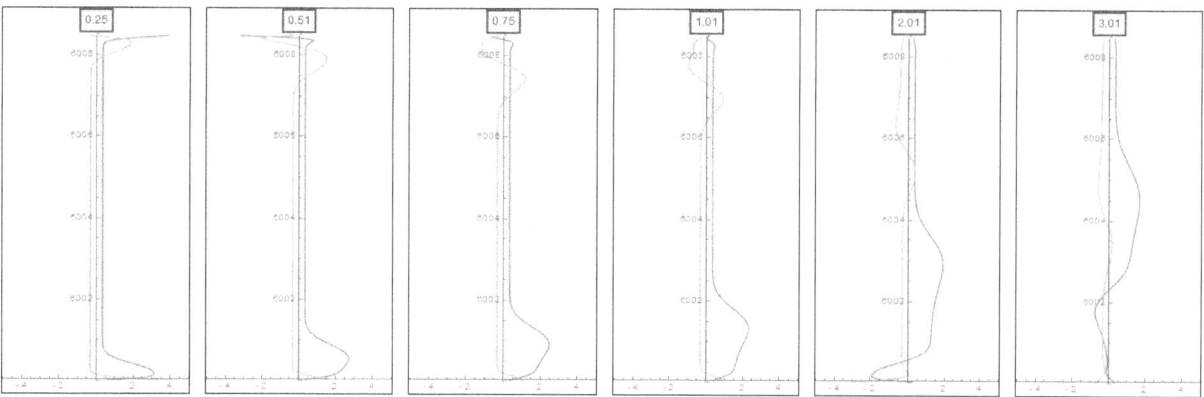

Figure 42 The first second shows the disturbance starting, and the next three seconds show the propagation

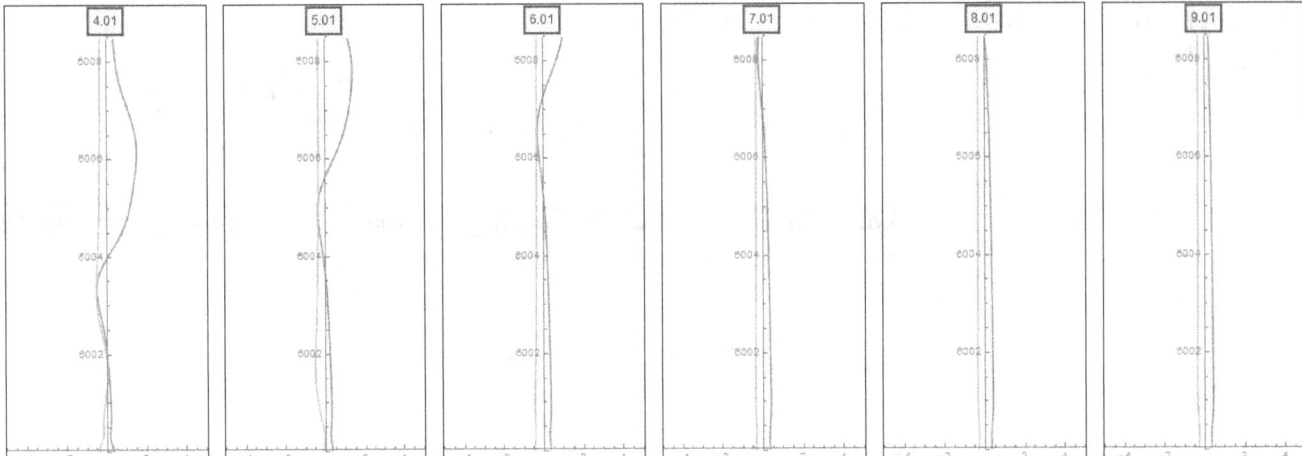

Figure 43 The remaining six seconds show the disturbance remaining within bounds, arriving at the other end and dissipating

In Figures 42 and 43, the disturbance is seen to propagate to the other end without increasing and eventually to die down.

The simulation was done using Wolfram Mathematica. It assumes that we can measure z_b and the relative lateral velocities $\frac{dz_b}{dt}$ and $\frac{\partial z_b}{\partial t}$ to calculate the expressions for R_b and Y in equations (2) and (3). More detail on how these quantities are measured and calculated is given in Appendix A. Once the force R_b has been calculated for each bolt, it is used directly to calculate the lateral acceleration and hence the velocity and displacement. After the disturbance has subsided, the damping and other forces become tiny, on the order of 10^{-6} Newtons. The only substantial force remaining is that provided by the permanent magnets to counteract the Coriolis force.

The equations of motion are not programmed into the simulation; the behavior they imply arises naturally from the application of Newton's laws of motion.

Mathematical details

The equation for a bolt's lateral acceleration [15] is

$$\frac{d^2z}{dt^2} = \frac{\partial^2z}{\partial t^2} + 2V\frac{\partial^2z}{\partial x\partial t} + V^2\frac{\partial^2z}{\partial x^2} \qquad (A.1)$$

The displacement along the direction of travel is x, the lateral displacement is z, the vertical velocity is V and time is t. Let R_a be the average force per unit distance on a stream of ascending bolts. Let the bolts' lateral displacement be z_a; their velocity is V and mass per meter is m. Then the equation of motion is

International Space Elevator Consortium

$$R_a = m\left(\frac{\partial^2 z_a}{\partial t^2} + 2V\frac{\partial^2 z_a}{\partial x \partial t} + V^2\frac{\partial^2 z_a}{\partial x^2}\right) \qquad (A.2)$$

Similarly for the second stream of bolts, in which the descending bolts' velocity is $-V$, the lateral displacement is z_d and the force per unit distance is R_d, the equation of motion is

$$R_d = m\left(\frac{\partial^2 z_d}{\partial t^2} - 2V\frac{\partial^2 z_d}{\partial x \partial t} + V^2\frac{\partial^2 z_d}{\partial x^2}\right) \qquad (A.3)$$

Between the streams of bolts, we can exert a force

$$R_b = R_a - R_d = m\left(\frac{\partial^2 z_b}{\partial t^2} + 4V\frac{\partial^2 z_s}{\partial x \partial t} + V^2\frac{\partial^2 z_b}{\partial x^2}\right) \qquad (A.4)$$

Here, $z_b = z_a - z_d$ is the distance between the bolts, and $z_s = (z_a + z_d)/2$ is the overall lateral displacement of the two streams of bolts together. An external lateral force R_e, due to the tidal effects of the sun, moon and planets, applies equally to both streams of bolts.

$$R_e = R_a + R_d = 2m\left(\frac{\partial^2 z_s}{\partial t^2} + V\frac{\partial^2 z_b}{\partial x \partial t} + V^2\frac{\partial^2 z_s}{\partial x^2}\right) \qquad (A.5)$$

A stable position exists in which the time derivatives are zero and the curvature satisfies

$$\frac{\partial^2 z_s}{\partial x^2} = \frac{R_e}{2mV^2} \qquad (A.6)$$

In words, there is a slight bowing of the streams of bolts as their paths curve in response to the external forces. The tides are wholly predictable, and the bolts need to be ejected from the ambit or the top of the first stage to align with the predicted curve.

The external force R_e is due to tides, and so it varies very slowly compared with the lengths and times that affect the force R_b between bolts. Therefore, its derivatives with respect to both x and t can be ignored. Hence and from equation (A.5) we can partially differentiate twice with respect to x to obtain, to a first approximation:

$$\frac{\partial^4 z_s}{\partial t^2 \partial x^2} + V\frac{\partial^4 z_b}{\partial x^3 \partial t} + V^2\frac{\partial^4 z_s}{\partial x^4} = \frac{\partial^2 R_e}{\partial x^2}/2m = 0 \qquad (A.7)$$

Similarly we partially differentiate equation (A.5) with respect to t:

$$\frac{\partial^4 z_s}{\partial t^4} + V\frac{\partial^4 z_b}{\partial x \partial t^3} + V^2\frac{\partial^4 z_s}{\partial x^2 \partial t^2} = \frac{\partial^2 R_e}{\partial t^2}/2m = 0 \qquad (A.8)$$

Differentiate equation (4) with respect to x and t.

$$\frac{\partial^2 R_b}{\partial x \partial t} = m\left(\frac{\partial^4 z_b}{\partial x \partial t^3} + 4V\frac{\partial^4 z_s}{\partial x^2 \partial t^2} + V^2\frac{\partial^4 z_b}{\partial x^3 \partial t}\right) \qquad (A.9)$$

Using equations (7) and (8), substituting for the two terms in z_b in equation (9).

$$V\frac{\partial^2 R_b}{\partial x \partial t} = m\left(-\left[\frac{\partial^4 z_s}{\partial t^4} + V^2\frac{\partial^4 z_s}{\partial x^2 \partial t^2}\right] + 4V^2\frac{\partial^4 z_s}{\partial x^2 \partial t^2} - \left[V^2\frac{\partial^4 z_s}{\partial x^2 \partial t^2} + V^4\frac{\partial^4 z_s}{\partial x^4}\right]\right) \qquad (A.10)$$

Therefore

$$V\frac{\partial^2 R_b}{\partial x \partial t} + m\left(\frac{\partial^4 z_s}{\partial t^4} - 2V^2\frac{\partial^4 z_s}{\partial x^2 \partial t^2} + V^4\frac{\partial^4 z_s}{\partial x^4}\right) = 0 \qquad (A.11)$$

This is equation (1) in the main text. From equation (A.11), we conclude that judicious control of the force R_b between bolts can control the overall lateral movement of the pair of streams.

Solution

Equation (A.11) relates R_b to a function of derivatives of z_s, so we seek a solution with suitable values of R_b in terms of z_s. We take the solution to be a linear sum of $z_s = e^{\omega t + \alpha x}$, where ω and α are complex in general. We then obtain the equation

$$F_b V \alpha \omega + m(\omega^4 - 2V^2 \alpha^2 \omega^2 + V^4 \alpha^4) = 0 \qquad (A.12)$$

Here, F_b is the Laplace transform of R_b. For stable damped harmonic motion in the gap of width z_b we need a restoring force of Az_b, a damping force $c\frac{\partial z_b}{\partial t}$, and some additional terms that we will designate Y:

$$R_b = m\left(Y - Az_b - c\frac{\partial z_b}{\partial t}\right) \qquad (A.13)$$

Equation (A.13) is equation (2) in the main text. Equation (A.11) requires a value $V\frac{\partial^2 R_b}{\partial x \partial t}$:

$$V\frac{\partial^2 R_b}{\partial x \partial t} = m\left(V\frac{\partial^2 Y}{\partial x \partial t} - AV\frac{\partial^2 z_b}{\partial x \partial t} - cV\frac{\partial^3 z_b}{\partial x \partial t^2}\right) \qquad (A.14)$$

Use equation (A.5) to substitute terms in z_s instead of z_b. We take R_e to be constant.

$$V\frac{\partial^2 R_b}{\partial x \partial t} = mV\frac{\partial^2 Y}{\partial x \partial t} - A\left[\frac{R_e}{2} - m\left(\frac{\partial^2 z_s}{\partial t^2} + V^2\frac{\partial^2 z_s}{\partial x^2}\right)\right] + mc\left(\frac{\partial^3 z_s}{\partial t^3} + V^2\frac{\partial^3 z_s}{\partial x^2 \partial t}\right)$$

As in equation (3) in the main text, we take

$$Y = m\left(\frac{l}{V}\int\frac{\partial z_s}{\partial t}dx + \frac{g}{V}\int z_s dx - 2cV\frac{\partial z_s}{\partial x} + \frac{h}{V}\iint z_s dx dt - 2AV\int\frac{\partial z_s}{\partial x}dt\right) \qquad (A.15)$$

Y has a Laplace transform of $m\left(\frac{l\omega}{V\alpha} + \frac{g}{V\alpha} - 2cV\alpha + \frac{h}{V\alpha\omega} - 2AV\frac{\alpha}{\omega}\right)$. The Laplace transform of $V\frac{\partial^2 R_b}{\partial x \partial t}$ is then as follows.

$$\begin{aligned}F_b V\alpha\omega &= m(l\omega^2 + (g - 2cV^2\alpha^2)\omega + h - 2AV^2\alpha^2) - A\left[\frac{1}{2}R_e - m(\omega^2 + V^2\alpha^2)\right] \\ &\quad + mc[\omega^3 + V^2\alpha^2\omega] \qquad (A.16)\end{aligned}$$

Therefore
$$F_b V\alpha\omega = m(c\omega^3 + (A + l)\omega^2 + (g - cV^2\alpha^2)\omega + h - A R_e/2m - AV^2\alpha^2) \quad (A.17)$$

Set $Q^2 = -A R_e/2m$. Then equation (A.12) gives us

$$\begin{aligned}\omega^4 + c\omega^3 + (A + l - 2V^2\alpha^2)\omega^2 + (g - cV^2\alpha^2)\omega + h + Q^2 - AV^2\alpha^2 + V^4\alpha^4 \\ = 0 \quad (A.18)\end{aligned}$$

If $c = l = g = 0$, equation (A.18) becomes

$$\omega^4 + (A - 2V^2\alpha^2)\omega^2 + h + Q^2 - AV^2\alpha^2 + V^4\alpha^4 = 0$$

It factorizes into a product of two terms as follows.

$$\left(\omega^2 - \frac{1}{2}\left[2V^2\alpha^2 - A \pm \sqrt{(2V^2\alpha^2 - A)^2 - 4(h + Q^2 - AV^2\alpha^2 + V^4\alpha^4)}\right]\right) = 0$$

This simplifies to the following.

$$\left(\omega^2 - \tfrac{1}{2}\left[2V^2\alpha^2 - A \pm \sqrt{A^2 - 4(Q^2 + h)}\right]\right) = 0$$

Write $W^2 = \tfrac{1}{4}A^2 - (Q^2 + h)$. Then the factorization is

$$\left(\omega^2 - \left[V^2\alpha^2 - \tfrac{1}{2}A + W\right]\right)\left(\omega^2 - \left[V^2\alpha^2 - \tfrac{1}{2}A - W\right]\right) = 0$$

The full factorization with g, l and c non-zero is

$$\left(\omega^2 - D\omega - \left[V^2\alpha^2 - \tfrac{1}{2}A + W\right]\right)\left(\omega^2 - E\omega - \left[V^2\alpha^2 - \tfrac{1}{2}A - W\right]\right) = 0$$

Comparing this with the coefficients of ω^3, ω^2 and ω in equation (A.18) we obtain three equations.

$$-D - E = c$$
$$DE = l$$
$$D\left[V^2\alpha^2 - \tfrac{1}{2}A + W\right] + E\left[V^2\alpha^2 - \tfrac{1}{2}A - W\right] = g - cV^2\alpha^2$$

Hence $D(-c - D) = l$, so $D^2 + cD + l = 0$, and $D = \tfrac{1}{2}\left[-c \pm \sqrt{c^2 - 4l}\right]$. Similarly $E = \tfrac{1}{2}\left[-c \mp \sqrt{c^2 - 4l}\right]$. Without loss of generality, we can deal with D alone because E is similar.

The third equation gives a choice of two values for g as follows.

$$-c\left(V^2\alpha^2 - \tfrac{1}{2}A\right) \pm W\sqrt{c^2 - 4l} = g - cV^2\alpha^2$$
$$g = \tfrac{1}{2}Ac \pm W\sqrt{c^2 - 4l}$$

The solution for ω includes the constants D and D^2.

$$\omega = \tfrac{1}{2}\left[D \pm \sqrt{D^2 + 4\left(V^2\alpha^2 - \tfrac{1}{2}A \pm W\right)}\right]$$
$$D = \tfrac{1}{2}\left[-c \pm \sqrt{c^2 - 4l}\right]$$
$$D^2 = \tfrac{1}{2}\left[c^2 - 2l \pm c\sqrt{c^2 - 4l}\right]$$

For ω to converge, we need the real part of D to be negative, i.e., $\mathbb{R}D = c/2 > 0$. We require $4l > c^2$ to make the square root in D imaginary. Now $Q^2 = A R_e/2m$ is real; it may be positive or negative. Next, $W = \sqrt{\frac{1}{4}A^2 - (Q^2 + h)}$. To make g real, and hence implementable, we need W to be imaginary. Therefore we need $h > \frac{1}{4}A^2 - Q^2$.

Then, in the expression for ω, the square root contains the imaginary terms $\pm 4W \pm c\sqrt{c^2 - 4l}$, which will give a real part. We need to show that it is less than $\mathbb{R}D = c/2$ to ensure that the sum of the real parts of ω is negative and therefore convergent. Let $\omega = \frac{1}{2}\left(D \pm \sqrt{Pe^\theta}\right) = \frac{1}{2}\left(D \pm \sqrt{P}e^{\theta/2}\right)$. The real part of $\sqrt{P}e^{\theta/2}$ is $\sqrt{P}\cos\frac{\theta}{2}$. We require $\frac{1}{2}c > \sqrt{P}\cos\frac{\theta}{2}$ or equivalently $c^2 > 4P\cos^2\frac{\theta}{2}$, i.e., $c^2 > 2P(1 + \cos\theta)$ or

$$[c^2 - 2P\cos\theta]^2 > 4P^2$$

That is:

$$c^4 - 4c^2 P\cos\theta > 4P^2\sin^2\theta$$

Now

$$P\cos\theta = \frac{1}{2}\{c^2 - 2l\} + 4\left\{V^2\alpha^2 - \frac{1}{2}A\right\}$$

and

$$iP\sin\theta = \pm 4W \pm \frac{1}{2}c\sqrt{c^2 - 4l}$$

So the required condition is as follows.

$$c^4 - 4c^2\left(\frac{1}{2}\{c^2 - 2l\} + 4\left\{V^2\alpha^2 - \frac{1}{2}A\right\}\right) > -4\left\{16W^2 \pm 4cW\sqrt{c^2 - 4l} + \frac{1}{4}c^2(c^2 - 4l)\right\}$$

Hence

$$c^4 - c^2\left(2\{c^2 - 2l\} + 16\left\{V^2\alpha^2 - \frac{1}{2}A\right\}\right) + c^2(c^2 - 4l) + 64W^2 > \pm 16cW\sqrt{c^2 - 4l}$$

Therefore

$$64W^2 + c^2\{8A - 16V^2\alpha^2\} > \pm 16cW\sqrt{c^2 - 4l}$$

Keeping l only a little higher than $\frac{1}{4}c^2$ and setting h to keep W small ensures that the RHS is small. On the LHS the dominant term is $-16c^2V^2\alpha^2 > 0$. $W^2 < 0$. The condition may be written

$$\frac{1}{2}A - V^2\alpha^2 > -4\frac{W^2}{c^2} \pm \frac{W}{c}\sqrt{c^2 - 4l}$$

International Space Elevator Consortium

All the terms are positive, assuming α is imaginary, but the RHS is small by design. To guarantee satisfaction for all wavelengths $\lambda = 2\pi i/\alpha$ we should make

$$A > -2\left[4\frac{W^2}{c^2} + \frac{W}{c}\sqrt{c^2 - 4l}\right]$$

Since W is proportional to A, this inequality is essentially a condition on c.

Using the solution

Combining equations (A.13) and (A.15) gives a solution

$$R_b = m\left(\frac{l}{V}\int\frac{\partial z_s}{\partial t}dx + \frac{g}{V}\int z_s dx + \frac{h}{V}\iint z_s dxdt - A\left[z_b + 2V\int\frac{\partial z_s}{\partial x}dt\right]\right.$$
$$\left. - c\left[\frac{\partial z_b}{\partial t} + 2V\frac{\partial z_s}{\partial x}\right]\right) (A.19)$$

This result for R_b requires $\frac{\partial z_s}{\partial x}, \frac{\partial z_s}{\partial t}$ and various integrals to be calculated for each bolt. To calculate $\frac{\partial z_s}{\partial x}$, consider the first-order equation that underlies equation (A.1) for an ascending bolt.

$$\frac{dz_a}{dt} = \frac{\partial z_a}{\partial t} + V\frac{\partial z_a}{\partial x}$$

Similarly, for a descending bolt,

$$\frac{dz_d}{dt} = \frac{\partial z_d}{\partial t} - V\frac{\partial z_d}{\partial x}$$

When a moving bolt measures z_b and then $\frac{dz_b}{dt}$, it obtains

$$\frac{dz_b}{dt} = \frac{dz_a}{dt} - \frac{dz_d}{dt} = \frac{\partial z_a}{\partial t} + V\frac{\partial z_a}{\partial x} - \left[\frac{\partial z_d}{\partial t} - V\frac{\partial z_d}{\partial x}\right] = \frac{\partial z_b}{\partial t} + 2V\frac{\partial z_s}{\partial x} \qquad (A.20)$$

This formula directly gives the term in c in equation (A.19).

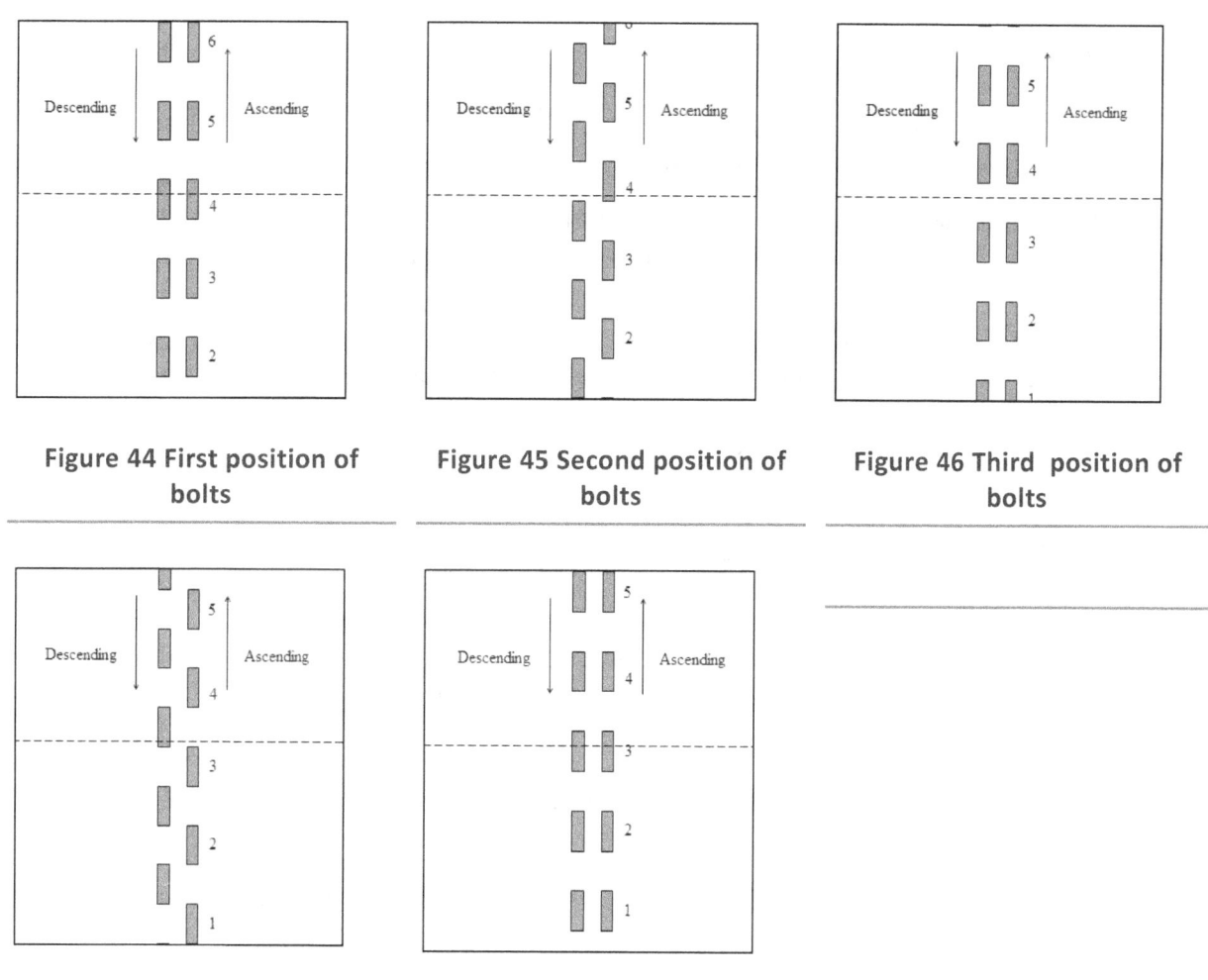

Figure 44 First position of bolts	Figure 45 Second position of bolts	Figure 46 Third position of bolts

Figure 47 Fourth position of bolts	Figure 48 Fifth position of bolts	

Now $\frac{\partial z_b}{\partial t}$ is the rate of change of lateral displacement between the two streams of bolts traveling in opposite directions at a particular position in space, whereas $\frac{dz_b}{dt}$ is the rate of change as seen by a moving bolt. To obtain $\frac{\partial z_b}{\partial t}$, at least two neighboring bolts must communicate. In Figure 44, bolt 4 is aligned with the fixed horizontal dashed line. As it ascends, the other stream of bolts descends. After a short time, bolt 3 reaches the position marked by the dashed line (Figure 48). To measure $\frac{\partial z_b}{\partial t}$ it needs to know the value of z_b recorded by bolt 4 when it was there. Bolt 4 has to pass this information to bolt 3, and so on between every pair of bolts. It may be desirable to average the value over more than two bolts, but the principle is the same.

International Space Elevator Consortium

Once the bolt has measured $\frac{dz_b}{dt}$ and $\frac{\partial z_b}{\partial t}$, it can use the formula (A.20) to calculate $\frac{\partial z_s}{\partial x}$. To obtain the term $A\left[z_b + 2V\int \frac{\partial z_s}{\partial x}dt\right]$ in equation (A.19) it can integrate $\left[\frac{\partial z_b}{\partial t} + 2V\frac{\partial z_s}{\partial x}\right]$ with respect to t by using a similar technique to that used to obtain $\frac{\partial z_b}{\partial t}$. Again, bolt 4 passes its accumulated value for the integral to bolt 3 so that bolt 3 can add its measurement to calculate the new integral giving a value at the same position, i.e., at the same value of x.

The terms in l, g and h require integrals with respect to x, which is very difficult to perform, as it involves simultaneously evaluating the positions of the bolts along the length of the stream at any moment in time t. A better solution is to use equation (A.5), from which we can obtain the following relationship by integrating with respect to x and t.

$$\iint R_e dx dt = 2m\left(\int \frac{\partial z_s}{\partial t}dx + Vz_b + V^2\int \frac{\partial z_s}{\partial x}dt\right) \qquad (A.21)$$

Hence, the term in l can be obtained in terms of a more manageable integral with respect to t, using the method already outlined above for the term in A.

$$m\frac{l}{V}\int \frac{\partial z_s}{\partial t}dx = \frac{1}{2}\frac{l}{V}\iint R_e dx dt - ml\left(z_b + V\int \frac{\partial z_s}{\partial x}dt\right) \qquad (A.22)$$

Similarly, the terms in g and h can be obtained by further integrating equation (A.22) with respect to t. The integrals of R_e can readily be calculated, because it is a slowly changing and wholly predictable quantity due to tidal forces.

One apparent source of difficulty is calculating the precise value of $V\int \frac{\partial z_s}{\partial x}dt$ required for the terms in A and l in equations (A.19) and (A.22). Fortunately, this is easy to deal with, because l does not need a precise value; it merely has to satisfy $l > \frac{1}{4}c^2$. Hence, a value of l somewhat larger than the minimum can compensate for the lack of precision in the A term. The value of g is exact according to the equations, where $g = \frac{1}{2}Ac \pm W\sqrt{c^2 - 4l}$. The effect of imprecision in g is explored numerically in the next section. The constant h is defined by the inequality $h > \frac{1}{4}A^2 - Q^2$, and so precision is not an issue.

There may be constants of integration where indefinite integrals are used. As we are treating R_e as constant, it makes sense that the constants of integration should cancel out the integrals of R_e. This matter may require further examination.

We do not want a constant component in R_b; the Coriolis force is taken care of by the straightforward action of the permanent magnets. The value of z_b implicitly compensates for this and it does not appear explicitly in the equations. The term Az_b in equation (A.19) reflects the action of the permanent magnets and does not require a current in the

International Space Elevator Consortium

electromagnetic coils. A current is needed for damping the other terms in equation (A.19), which are very small once a stable state has been reached. A current is also needed for maintaining stability when one bolt's arm is positioned between two arms of a bolt traveling in the opposite direction.

Numerical analysis

The requirement for convergence is that the real part of ω in $e^{\omega t + \alpha x}$ should be negative so that any perturbations in the movement of the streams of bolts will die down over time. As ω satisfies a quartic equation, the following four graphics (Figures 49 through 52) show the value of the real part of ω as a series of surfaces plotted against the values of V and α/i expected up to an altitude of 6000 km. Each graphic has five surfaces for different values of g at 0.8, 0.9, 1.0, 1.1 and 1.2 times the value $g = \frac{1}{2}Ac + W\sqrt{c^2 - 4l}$ that appears in the above analysis. This confirms that there is reasonable flexibility in making the measurements required for the term in g in equation (A.19).

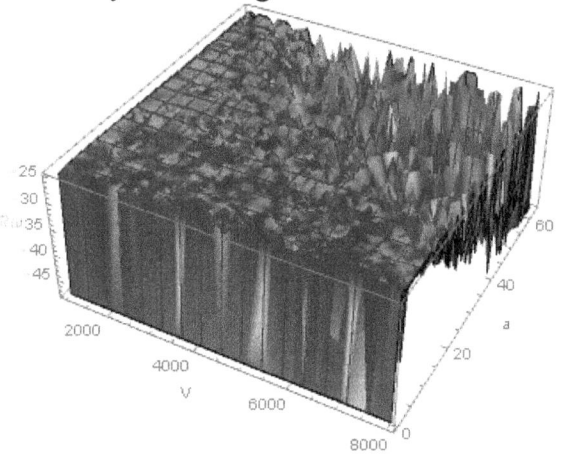

Figure 49 Plot of the real part of ω (1st solution) with g varying ±20%

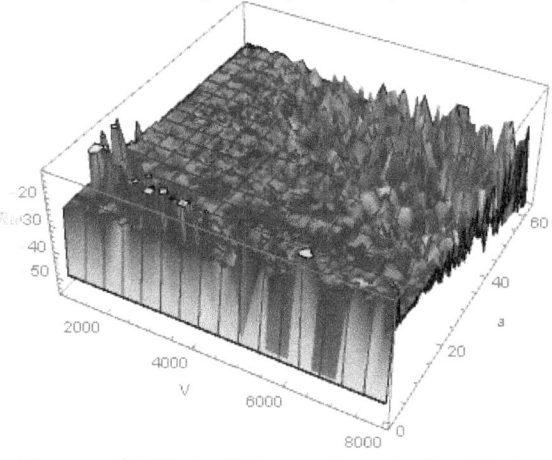

Figure 50 Plot of the real part of ω (2nd solution) with g varying ±20%

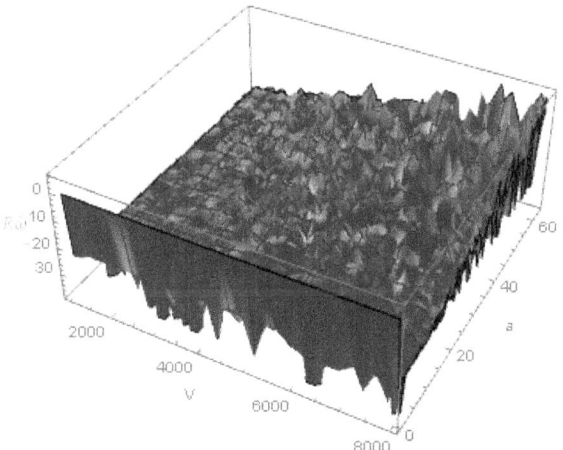

Figure 51 Plot of the real part of ω (3rd solution) with g varying ±20%

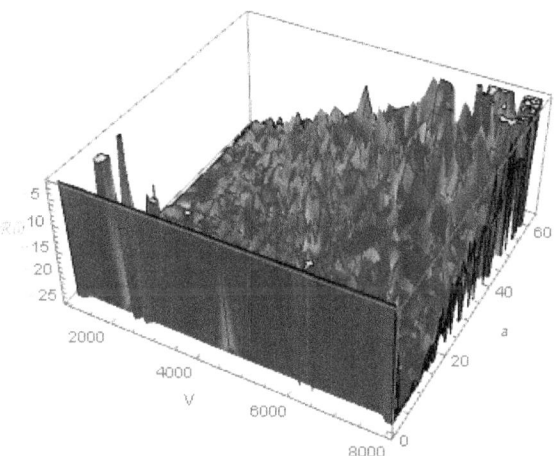

Figure 52 Plot of the real part of ω (4th solution) with g varying ±20%

International Space Elevator Consortium